ALARMING!
THE CHASM SEPARATING BASIC STATISTICS EDUCATION FROM ITS NECESSITIES

Copyright © 2013 by William J. Adams.

Library of Congress Control Number: 2012909643
ISBN: Hardcover 978-1-4771-2012-5
Softcover 978-1-4771-2011-8
Ebook 978-1-4771-2013-2

All rights reserved. No part of this book may be reproduced or transmitted in any form or by any means, electronic or mechanical, including photocopying, recording, or by any information storage and retrieval system, without permission in writing from the copyright owner.

This book was printed in the United States of America.

Rev. Date: 05/06/2013

To order additional copies of this book, contact:
Xlibris Corporation
1-888-795-4274
www.Xlibris.com
Orders@Xlibris.com
99964

ALARMING!
THE CHASM SEPARATING
BASIC STATISTICS EDUCATION
FROM ITS NECESSITIES

Number Pushing and Glitz
are Insufficient for
Statistics Education

WILLIAM J. ADAMS
Mathematics Department
Pace University

with illustrations by
Ramunė B. Adams

To Sabrina

Acknowledgment: I am indebted to my daughter Ramunė for preparing the illustrations.

Preface

The fairly large sample of current basic statistics books I gave thought to recently may, in my view, be characterized as number pushers with a large number of illustrations intended to convey a sense of the importance of statistics to the study of real-world problems.

What's wrong with that? Nothing, provided that what I submit to be the necessities of statistics education are given the attention they warrant, are not smothered by glitz, overwhelming attention to number pushing, do not receive shortshrift, or are not mentioned at all.

Am I being too critical? I invite you to give thought to fifteen, issues/questions that are the core of the aforenoted chasm and render your verdict.

Food-for-thought in support of the issues/questions raised along with answers/discussion are included.

Many students and those who apply statistics to their fields of interest subscribe to the view that to apply statistics to a problem/situation all you need do is throw your data into a computer and let it do its thing. The further they travel along this road of thought the more we can expect misunderstanding of statistics and, when it comes to publication, statistical junk.

I believe that the only way to change the direction of this road of thought is to incorporate into our teaching of basic statistics what is feasible of the necessities of statistics education.

For further discussion of the necessities of basic statistics education I recommend the book I coauthored with my colleagues Irwin Kabus and Mitchell Preiss: *Statistics: Basic Principles and Applications*, 2nd ed, (Kendall/Hunt Publishing Co., 2000) or the revised 2nd ed., W. J. Adams (Xlibris, 2009).

W. J. A.

Brief Contents

Preface

Contents

1   Strings Attached That Underlie the Application of Statistical Methods? Yes, Ignore Them at Your Peril.

2   If the Strings Attached are Detached in an Application, What Then? Robustness to the Rescue?

3   If the Strings Attached are Detached in an Application and Robustness Fails, What Then?

4   Good Data, Good Data, My Kingdom for Good Data (Part 1): Is Random Sampling in Practice as Simple as it Sounds in Theory?

5   Good Data, Good Data, My Kingdom for Good Data (Part 2): How Reliable are These Data? How Relevant are They to the Situation Under Study?

6   Good Data, Good Data, My Kingdom for Good Data (Part 3): Can You Trust Polls?

7   These Numbers/Statistics are Not Reliable or Not Relevant. So What?

8   Are Quantitative Studies Preferable to Qualitative Ones? Can They Be Combined?

9   You Want More Accuracy in Your Figures? Add More Decimals to Them?

10   Statistics Tea, Leaves: What Do These Statistics Tell Us? Their Limitations?

11   Data Scales: Nominal, Ordinal, Interval, Ratio: Caution Advised.

12   A Role for Mathematical Modeling in Teaching Basic Statistics? Yes, Most Important.

13   KISS (Keep It Simple Stupid), But Keep It Correct.

14   In Teaching Basic Statistics "Realistic" Data Should Be Used, or Not Necessarily?

15   I Believe That Computers and Statistical Packages Should NOT Be Used in Teaching Basic Statistics. Heretic, Lunatic, Luddite, or Not?

Tables

Index

CONTENTS

Preface..9
Brief Contents..11

1 Strings Attached That Underlie the Application of Statistical Methods? Yes, Ignore Them at Your Peril.................................. 17
   1.1 Preface..17
   1.2 The Small Sample Size Confidence Interval for the Population Mean..19
   1.3 Food for Thought..20
   1.4 Answers / Discussion of Food for Thought.............22
   1.5 Estimation of Variance and Standard Deviation.....24
   1.6 Food for Thought..24
   1.7 Answers/Discussion of Food for Thought..............26
   1.8 Hypothesis Testing for the Mean of a Population: Small Sample Size 27
   1.9 Food for Thought..29
   1.10 Answers/Discussion of Food for Thought............30
   1.11 Hypothesis Testing for Equality of Variances and Standard Deviations..31
   1.12 Food for Thought..32
   1.13 Answers/Discussion of Food for Thought............33

2 If the Strings Attached are Detached in an Application, What Then? Robustness to the Rescue? 35
   2.1 Preface..35
   2.2 Robustness 35
   2.3 The Lilliefors Test..36
   2.4 Food for Thought..39
   2.5 Answers /Discussion of Food for Thought.............40

3	**If the Strings Attached are Detached in an Application and Robustness Fails, What Then?** 42	
	3.1 Preface...42	
	3.2 Mathematics Review Tapes 42	
	3.3 A Paired-Sample Sign Test 43	
	3.4 Return to the Mathematics Review Tape Situation 44	
	3.5 Food for Thought..46	
	3.6 Answers /Discussion of Food for Thought 48	

4	**Good Data, Good Data, My Kingdom for Good Data (Part 1): Is Random Sampling in Practice as Simple as It Sounds in Theory?** 50	
	4.1 Preface...50	
	4.2 Can We Trust TV Ratings? 51	
	4.3 Going to War 52	
	4.4 Randomness Achieved in Practice? Hypothesis Tests Say No, Reality Agrees. 54	
	4.5 Food for Thought..55	
	4.6 Answers/Discussion of Food for Thought 56	

5	**Good Data, Good Data, My Kingdom for Good Data (Part 2): How Reliable are These Data? How Relevant are They to the Situation Under Study?** 57	
	5.1 Preface...57	
	5.2 Are the Data Reliable? 58	
	5.3 Food for Thought..71	
	5.4 Answers/Discussion of Food for Thought 74	
	5.5 Are the Data Relevant to the Situation Under Consideration? 79	
	5.6 Food for Thought..85	
	5.7 Answers / Discussion of Food for Thought 86	

6	**Good Data, Good Data My Kingdom for Good Data (Part 3): Can You Trust Polls?** 89	
	6.1 Preface...89	
	6.2 How Trustworthy are These Poll Results? 90	
	6.3 Food for Thought..92	
	6.4 Answers/Discussion of Food for Thought 95	

7	These Numbers / Statistics are Not Reliable or Not Relevant. So What?	98
	7.1 Preface	98
	7.2 Cases	98

8	Are Quantitative Studies Preferable to Qualitative Ones? Can They Be Combined?	111
	8.1 Preface	111
	8.2 Sexuality By the Numbers, or Not?	111

9	You Want More Accuracy in Your Figures? Add More Decimals to Them?	113
	9.1 Preface	113
	9.2 Food for Thought	114
	9.3 Answers/Discussion of Food for Thought	115

10	Statistics Tea Leaves: What Do These Statistics Tell Us? Their Limitations?	118
	10.1 Preface	118
	10.2 Cases	119
	10.3 Food for Thought	125
	10.4 Answers/Discussion of Food for Thought	126

11	Data Scales: Nominal, Ordinal, Interval, Ratio: Caution Advised.	129
	11.1 Preface	129
	11.2 Data Scales	129
	11.3 Food for Thought	134
	11.4 Answers/Discussion of Food for Thought	136

12	A Role for Mathematical Modeling in Teaching Basic Statistics ? Yes, Most Important.	139
	12.1 Preface	139
	12.2 Model 1: Testing for the Difference Between Means, Small Samples, Independently Chosen, Approach	140
	12.3 Food for Thought	141
	12.4 Answers/Discussion of Food for Thought	144

	12.5	Model 2: Test Statistic for The Matched Pairs' Mean Difference Approach 150
	12.6	Food for Thought..154
	12.7	Answers/Discussion of Food for Thought.........................157
	12.8	The Consumer Price Index (CPI) Model..........................159

13	KISS (Keep It Simple Stupid), But Keep It Correct. 165	
	13.1	Preface..165
	13.2	"At Least" and "At Most" Claims 166
	13.3	An Incorrect Formulation of the Null Hypothesis for the "At Least" Claim..168
	13.4	Large Sample Size, n=30, Irrespective of the Nature of the Population? 169

14	In Teaching Basic Statistics "Realistic" Data Should Be Used, or Not Necessarily? 171	
	14.1	Preface..171
	14.2	Food for Thought..171
	14.3	Answers/Discussion of Food for Thought.........................173

15	I Believe That Computers and Statistical Packages Should Not Be Used in Teaching Basic Statistics: Heretic, Lunatic, Luddite, or Not? 174	
	15.1	Preface..174
	15.2	W.J. Adams et al; Concerning the Use of Computers and Statistical Packages in a Basic Level Statistics Course 175
	15.3	Excel Fans: CAUTION 181

Table

A	Standard Normal Curve 182
B	t Distribution 184
C	Chi-Square Curve..186
D	$F_{.05}(v_1, v_2)$...188
E	$F_{.01}(v_1, v_2)$...189
F	Lilliefors Test Bounds 190

Index..191

1

Strings Attached That Underlie the Application of Statistical Methods? Yes, Ignore Them at Your Peril.

1.1 PREFACE

With the emphasis on number pushing that has come to dominate the basic statistics scene many, dare I say most, students and users of statistics, have come to view it as a calculation free-for-all.

To counter this view conditions (I call them strings attached) must be given a prominent role in the study of statistics and reinforced by what I term food-for-thought.

Perhaps the best way to make the point about the significance of strings attached is to give an example about the price we pay when the strings attached are not satisfied. In their discussion of robustness in hypothesis testing Hoel, Port, and Stone[1] take up a hypothesis test employing a chi-square variable to test the null hypothesis

$$H_o : \sigma^2 = \sigma_o^2$$

to illustrate a lack of robustness with respect to the assumed normality of the underlying population. Hoel, Port, and Stone consider the underlying population distributed according to

$$y = xe^{-x}, x > 0$$

(see Figure 1.1). They show that the

---

[1] P. Hoel, S. Port, C. Stone, *Introduction to Statistical Theory* (Boston: Houghton Mifflin, 1971)

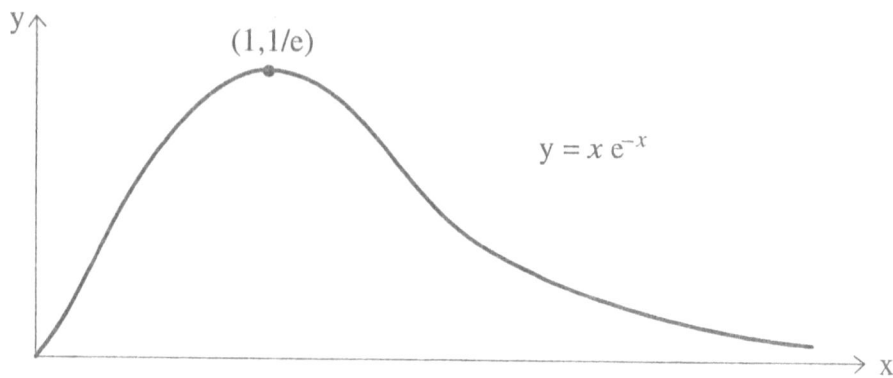

Figure 1.1

accept/reserve judgment region is approximately 1.58 times as long as the one obtained under the incorrect assumption (string attached) of normality or approximate normality of the population, so that the afore null hypothesis versus the two sided alternative would be rejected much more often than need be.

This example could be modified by removing the hypothesis testing framework to make it accessible after the concept of normality of a population has been introduced.

A Sample is Chosen at Random

In the sections that follow in this chapter and throughout statistics in general we see this statement and our general reaction is, so, "no big deal." It is far from being "no big deal," as is discussed in Chapter 4.

A sample is chosen at random should be regarded as a string attached to be realized in practice and not be taken for granted.

1.2 THE SMALL SAMPLE SIZE CONFIDENCE INTERVAL FOR THE POPULATION MEAN

> Confidence intervals for µ, where the sample size n is small, the sample is chosen at random from a population which is normally distributed or approximately so and the population standard deviation σ is not known, are based on the probability behavior of the sample statistic
>
> $$t = \frac{\bar{x} - \mu}{s/\sqrt{n}}$$
>
> where $\bar{x}$ is the sample mean. The variability of t is due to $\bar{x}$ and s.
>
> **Sampling Distribution of t.** Under the above assumptions (strings attached) the sampling distribution of
>
> $$t = \frac{\bar{x} - \mu}{s/\sqrt{n}}$$
>
> is approximately the t distribution with $n - 1$ degrees of freedom.

> **Strings Attached**
>
> Two strings attached for the application of the small sample confidence interval and error term are that the population being sampled is normally distributed or approximately so, and that the population standard deviation σ is not known. Roughly put, the first string means that a mound shaped population like that shown in Figure 1.2 may be sampled from, whereas highly skewed populations like those shown in Figures 1.3 and 1.4 do not satisfy the string attached.

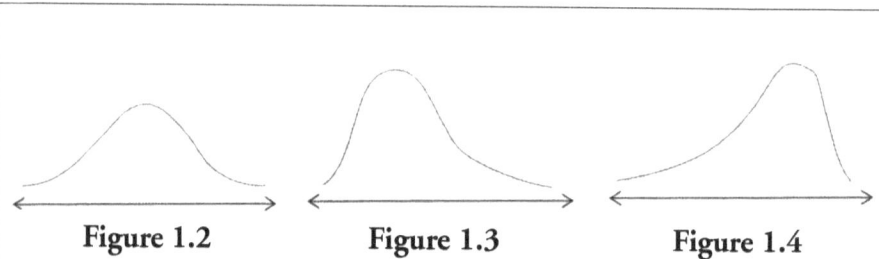

Figure 1.2   Figure 1.3   Figure 1.4

Why all this fuss about strings attached? Strings attached are like the small print in a legal document which we tend not to look at closely, often much to our subsequent dismay. When the strings attached to a theorem, test, or procedure are violated, the results promised are illusionary.

1.3   FOOD FOR THOUGHT

1. **Friendly Furry Creature**

    Robert Franks, an experimental psychologist, seeks an estimate of the time it takes a species of friendly furry creature he is studying to run through a maze. A randomly chosen sample of 10 friendly furry creatures traversed the maze in the following times, in seconds: 52, 48, 59, 60, 55, 51, 51, 50, 49, 58.

    (a) Determine the sample mean.

    (b) If the sample mean is taken as an estimate of the population mean, what is the error inherent in the estimate at the 0.95 confidence level?

    (c) How is the result obtained in answer to (a) to be interpreted?

    (d) What is the population whose mean Robert is seeking to estimate?

    (e) Is there a string attached on this population that Robert should be concerned with?

(f) Is there any other string attached Robert should be concerned about?

(g) If these strings attached are not met (detached), then what should be said about the reliability of the error estimate given in answer to (b) when the sample mean is taken as an estimate for the population mean?

2. **Jennifer Clarke**

Jennifer Clarke, a quality control manager for an auto manufacturer, wants to get an estimate for the mean dollar damage to the new Turbo model when it is driven into heavier autos at 30 miles per hour. She has been authorized to conduct tests with 5 Turbo model cars. A random sample of 5 cars that were tested yielded a mean damage amount of $18,250, with a standard deviation of $3500.

(a) Find 98% confidence limits for the population mean. How is your result to be interpreted?

(b) Evidence available for models similar to the Turbo suggest that the population of dollar damage values is U shaped, with values concentrated at the low and high ends, indicating that collision costs tended to be minimal or very high. Would this information prompt you to question the reliability of the confidence limits obtained in answer to (a)?

3. **The Tobin Company**

A test of the breaking strengths of 15 safety belts chosen at random from the production line of the Tobin Company yielded $\Sigma x_i = 1430$ (pounds) and $\Sigma x_i^2 = 137,200$ (pounds), where x_i is the breaking strength, in pounds, of the ith belt in the sample, the sum being taken from 1 to 15.

(a) Find 98% confidence limits for the mean breaking strength of the belts made by the Tobin Company.

(b) How is the result obtained in answer to (a) to be interpreted?

(c) If the sample mean is taken as an estimate for the population mean, what is the error inherent in the estimate at the 0.99 confidence level?

(d) How is the answer obtained in answer to (c) to be interpreted?

(e) What strings attached underlie the confidence limits and error term obtained in answer to (a) and (c)?

(f) Suppose it were later found that the population from which the sample was chosen deviates considerably from the a string attached given in answer to (e) or that the random sampling envisioned was not realized in practice; where does this leave us in connection with the result obtained in response to (b) and (d)?

1.4 Answers / Discussion of Food for Thought

1. **Friendly Furry Creature**

   (a) $\Sigma x = 533$; thus $\bar{x} = 53$

   (b) $\Sigma x^2 = 28,581$, and $s^2 = \dfrac{10(28,581) - (533)^2}{90} = 19.1$; $s = 4.4$

   $t_{.025} = 2.262$ for d.f. $= 9$

   $E = 2.262 \dfrac{(4.4)}{\sqrt{10}} = 3.1$ (sec.) with prob. 0.95

   (c) 0.95 expresses the reliability of the estimate. It is a numerical measure of our degree of belief that the error does not exceed 3.1 (seconds) if μ is estimated as 53 (seconds).

   (d) The population of times that it takes for the members of the species of friendly furry creature to run through the maze.

(e) The population must be normally distributed or approximately so. The sample chosen must be chosen without bias which favors certain samples being chosen over others.

(f) The error term would have to be considered unreliable in the sense that 0.95, which expresses the reliability of the estimate of μ by $\bar{x} = 53$, would be open to question.

2. **Jennifer Clarke**

(a) $18{,}250 \pm 3.747(3{,}500/\sqrt{5}) = 18{,}250 \pm 5{,}865;\ 12{,}385 < \mu < 24{,}115$ with prob. 0.98.

(b) Yes, the confidence limits are not reliable since the string attached to population normality or approximate normality is violated.

3. **The Tobin Company**

$\bar{x} = 95.3,\ s = 7.9$.

(a) $95.3 \pm 2.624(7.9/\sqrt{15}) = 95.3 \pm 5.35;\ 90 < \mu < 101$ with prob. 0.98

(b) 0.98 is a numerical measure of our degree of belief that μ is between 90 and 101.

(c) $E = 2.977(7.9/\sqrt{15}) = 6$ with prob. 0.99.

(d) 0.99 is a numerical measure of our degree of belief that the error does not exceed 6 pounds if μ is estimated as 95.3 (pounds).

(e) The normality or approximate normality of the underlying population; the randomness of the selection of the sample.

(f) The results would be unreliable because the normality string attached on the population or the assumed randomness of the sample selection is violated.

1.5 Estimation of Variance and Standard Deviation

Confidence intervals for σ^2 and σ, where the sample of size n is chosen at random from a population which is normally distributed, or "very close" to it, is based on the probability behavior of the sample statistic

$$\chi^2 = \frac{(n-1)s^2}{\sigma^2},$$

where s^2 is the sample variance and σ^2 is the population variance. This sample statistic is denoted by χ^2 because its sampling distribution is, to a close approximation, a chi-square distribution.

Sampling Distribution of χ^2: Under the afore strings attached the sampling distribution of

$$\chi^2 = \frac{(n-1)s^2}{\sigma^2},$$

is approximately the χ^2 distribution with $n - 1$ degrees of freedom.

1.6 Food for Thought

1. **Ramunė's Gourmet Coffee**

    Ramunė's Gourmet Coffee is negotiating a contract with a processing company to fill 10-ounce jars with a new coffee blend to be introduced into the market. Ramunė's Gourmet Coffee is concerned with the variability of the fill weights, which must be reasonably small, and the consulting firm they hired wants to obtain 98% confidence intervals for the variance and standard deviation of the coffee fills. A randomly chosen sample of 9 10-ounce jars yielded a variance of 0.42 (ounces).

    (a) Determine the 98% confidence interval for σ.

(b) How is this result to be interpreted?

(c) What is the significance of this result to the Ramunė Company's management.

(d) Are there strings attached for the afore result to be considered reliable?

(e) If either of these strings attached is not satisfied, what may we conclude?

2. **Expenses Charged**

In reviewing the expenses charged to Ramunė's Gourmet Coffee by its consulting firm the accountant noted a charge for work done to establish that the population of fill weights is normally distributed. He could not understand the necessity for this work and concluded that the consulting firm was adding some unnecessary work to the project in order to inflate the bill. If he came to you for an explanation of why this work was necessary, what would you tell him?

3. **The Russell Coal Company**

The management of the Russell Coal Company is interested in obtaining an estimate of the variation in the heat producing capacity of coal from a newly opened mine (in millions of calories per ton). A random sample of 6 specimens from the mine yielded a standard deviation of 192 (million-calories per ton).

(a) Determine the 95% confidence interval for the standard deviation of the mine's heat producing capacity.

(b) How is this result to be interpreted?

(c) It was found that the way in which the sample was chosen was not at random, no question about it. What significance, if any, does this have for your answers to (a) and (b)?

1.7 Answers/Discussion of Food for Thought

1. **Ramunė's Gourmet Coffee**

   (a) Since d.f. = 8, from Table C we obtain $\chi^2_{.01} = 20.090$ and $\chi^2_{.99} = 1.646$. This yields:

   $$\frac{8(0.42)}{20.090} < \sigma^2 < \frac{8(0.42)}{1.646} \text{ with prob. } 0.98$$

   $0.17 < \sigma^2 < 2.04$ with prob. 0.98
   Taking square roots gives us:
   $0.41 < \sigma < 1.43$ with prob. 0.98

   (b) The odds are 49 to 1 that the standard deviation of the coffee fills is between 0.41 (ounces) and 1.43 (ounces).

   (c) Ramunė's management will have to decide whether this is satisfactory or to insist that the processing company reset its filling machines.

   (d) The strings attached are that (a) the population of 10-ounce jar coffee weights are normally distributed, or approximately so, and (b) the sample is chosen at random - in an unbiased way that does not favor certain jars being chosen over others.

   (e) The statements in (a) and (b) are not reliable. If strings attached are not satisfied, the valid conclusions drawn on their basis do not hold.

2. **Expenses Charged**

   The work was necessary because the normality, or approximate normality, of the 10-ounce fill weights is a string attached for the validity of the confidence interval that was obtained.

3. **Russell Coal Company**

(a) $\sqrt{\dfrac{5(192)^2}{12.8325}} < \sigma < \sqrt{\dfrac{5(192)^2}{0.831211}}$; $120 < \sigma < 471$ with prob. 0.95

(b) 0.95 is a numerical measure of our degree of belief that σ is between 120 and 471 or equivalently, the odds are 19 to 1 that σ is between 120 and 471.

(c) If a string attached is not satisfied for the valid conclusion in (a) and its interpretation in (b), then the result in (a) and its interpretation in (b) are not reliable and are open to question.

1.8 Hypothesis Testing for the Mean of a Population: Small Sample Size

1. **Ramunė's Gourmet Coffee**

    The management of Ramunė's Gourmet Coffee (RGC) decided to carry out a hypothesis test on the claim that the mean weight of their 8 ounce jars of the Deluxe Blend is 8 ounces, as stated

    (a) Formulate null and alternative hypothesis concerning this claim.

    (b) What is the basis for your alternative hypothesis?

    (c) Take $\alpha = 0.05$ as your level of significance.

    (d) It has been recommended to the management team that a sample size of 15 be used since small sample sizes are less costly to work with. Would you agree that this is the only consideration that favors small samples?

    (e) Is there any string attached that RGC's management should be concerned about?

(f) It has been decided to take a sample size of 15 to carry out the hypothesis test. What test statistic should be employed? What are the accept/reserve judgment and reject regions?

(g) The sample yielded the mean $\bar{x}$ = 8.2 (ounces) and standard deviation s = 0.8 (ounces). What significance does this have for RGC?

(h) It was determined that the analysis conducted which led to the conclusion that the population of fill weights from which the sample is drawn is approximately normal was faulty and that the population is U shaped. What significance, if any, does this have for RGC?

Discussion

1. **Ramunė's Gourmet Coffee**

   (a) $H_o: \mu = 8$

   $H_a: \mu \neq 8$

   (b) RGC is concerned with both overfills ($\mu > 8$), which are costly because more of the product is being given away than expected, and underfills ($\mu < 8$) which means that the company's product is not living up to expectations and the consumer is being short-changed. This situation may draw negative reactions from consumers and consumer protection groups and, needless-to-say, is not good for the company's image and business.

   (c) $\alpha = 0.05$

   (d) Less costly is desirable, but the basic issue is the normality or approximate normality of the population of fill weights. This string attached should be the focus of management's attention.

   (e) The randomness of the sample's selection, a string attached, must be realized in practice.

(f) $t = \dfrac{\bar{x} - 8}{s/\sqrt{15}}$ See Figure 1.5.

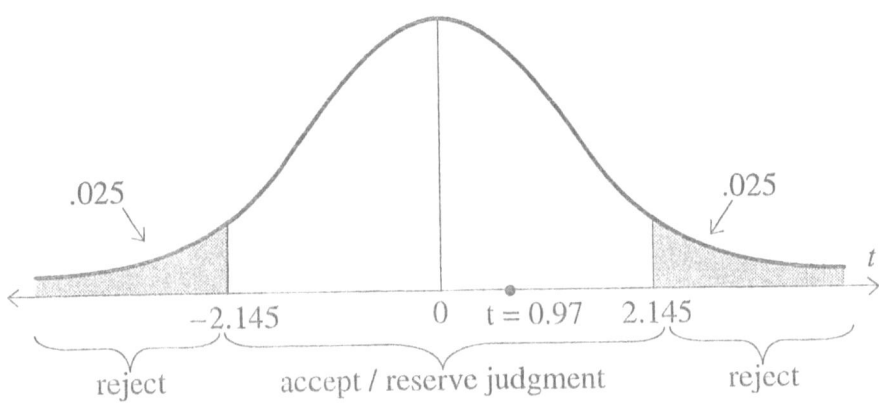

Figure 1.5

(g) $t = \dfrac{8.2 - 8}{0.8/\sqrt{15}} = 0.97$, which falls in the accept/reserve judgment region.

There are no constraints on RGC to not accept the null hypothesis and continue production.

With all such situations we should be aware of the possibility of a Type II error.

(h) The conclusion stated in (g) is open to question as is RGC's interpretation of this conclusion.

1.9 FOOD FOR THOUGHT

1. **The Midwest Economic Development Association (MEDA)**

   A group of labor economists has put out a report claiming that the mean income of factory workers in the midwest does not exceed $25,000. The Midwest Economic Development Association (MEDA) desires to test this claim. The 0.01 level of significance is to be used.

(a) Formulate null and alternative hypotheses.

(b) What is the basis for your null and alternative hypotheses?

(c) It has been suggested that a sample of size 10 be chosen because it is relatively easy to work with, does not consume an inordinate amount of time, and allows us to stay within budget. What is your reaction to this suggestion?

(d) Suppose that a sample of size 10 were to be used. What would the accept/reserve judgment and reject regions be?

(e) The analysis team ultimately decided on a sample size of 35. Could the same accept/reserve judgment and reject region you described in answer to (d) be used in this case?

(f) If you answered no to (e), then what accept/reserve judgment and reject regions would you recommend?

(g) Let us suppose that a random sample of size 35 yielded $\bar{x}$ = 25,300 (dollars) with s = 800 (dollars). What conclusion do you reach from these data?

(h) Do you have any concern about (g)?

(i) What does your finding mean to MEDA in terms of possible follow-up action?

1.10 Answers/Discussion of Food for Thought

1. **The Midwest Economic Development Association (MEDA)**

    (a) H_o: μ = 25,000. H_a: μ > 25,000.

    (b) H_a states the only possibility contradicting the claim.

(c) I would be concerned with the normality or approximate normality of the underlying population.

(d) $t = \dfrac{\bar{x} - 25{,}000}{s/\sqrt{10}}$. Accept/r. j. if $t \le 2.82$. Reject if $t > 2.82$.

(e) No; for large sample size the analysis is based on the z-bound 2.33.

(f) $z = \dfrac{\bar{x} - 25{,}000}{\sigma/\sqrt{35}}$. Accept/r. j. if $z \le 2.33$. Reject if z 2.33.

(g) $z = 2.22 \le 2.33$. Accept/r. j. on H_o.

(h) The random sampling string attached be realized in practice.

(i) They have no basis for taking issue with the claim. No follow-up action is indicated.

1.11 HYPOTHESIS TESTING FOR EQUALITY OF VARIANCES AND STANDARD DEVIATIONS

Sampling Distribution of $F = \dfrac{s_1^2/\sigma_1^2}{s_2^2/\sigma_2^2}$

It can be shown that if samples of size n_1 and n_2 are randomly chosen from normally distributed populations, or approximately so, with variances σ_1^2 and σ_2^2, the sampling distribution of the variance ratio

$$F = \dfrac{s_1^2/\sigma_1^2}{s_2^2/\sigma_2^2}$$

may be approximated by the F distribution with $v_1 = n_1 - 1$ and $v_2 = n_2 - 1$ degrees of freedom.

As the notation itself suggests, s_1^2 and s_2^2 are the sample variances of the samples drawn from the populations with variances σ_1^2 and σ_2^2, respectively.

With respect to the null hypothesis of variance equality, the test statistic reduces to the ratio of the sample variances:

$$F = \frac{s_1^2}{s_2^2},\ v_1 = n_1 - 1,\ v_2 = n_2 - 1$$

1.12 Food for Thought

1. **Variance Equality**

   Two populations P_1 and P_2 are, to a close approximation, normally distributed. Samples of size 11 are drawn at random from P_1 and P_2. They have variances $s_1^2 = 5.8$ and $s_2^2 = 4.2$, respectively. Test the hypothesis of variance equality at the 0.10 level.

   Questions

2. Consider two normally distributed populations with equal variances. Let us suppose that a hypothesis test for equality of variances is to be carried out versus the alternative hypothesis of unequal variances at the 0.02 level.

   (a) Show that if samples of size 10 are chosen at random from the populations, one sample variance could be more than five times the other without leading to rejection of the null hypothesis of equal population variances.

   (b) Show that if samples of sizes 5 are chosen at random from the populations, one sample variance could be almost sixteen times the other without leading to rejection of the null hypothesis of equal population variances.

   (c) What are the counterparts of the aforenoted factors 5 and 16 for the 0.10 significance level?

   (d) What do these results suggest concerning hypothesis testing for variance equality by means of small samples?

1.13 ANSWERS/DISCUSSION OF FOOD FOR THOUGHT

1. **Variance Equality**

   $v_1 = 10$ and $v_2 = 10$. Thus, we have:

   $$F_{.05}(10, 10) = 2.98, \quad F_{.95}(10, 10) = 1/2.98 = 0.34$$

   This gives us the bounds shown in Figure 1.6.

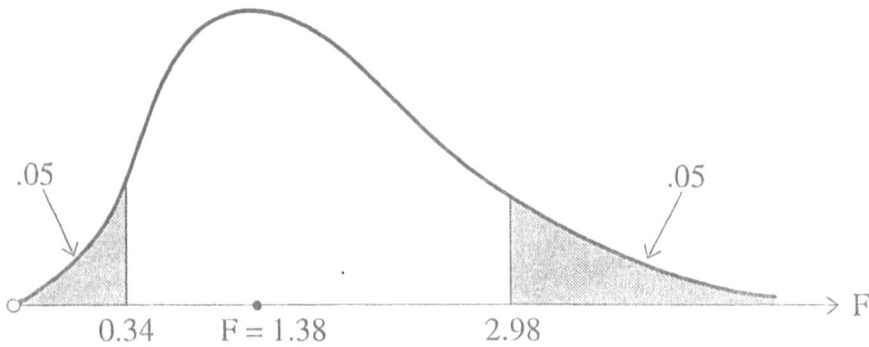

Figure 1.6

$s_1^2 = 5.8$ and $s_2^2 = 4.2$

so that

$$F = \frac{s_1^2}{s_2^2} = \frac{5.8}{4.2} = 1.38$$

Since F = 1.38 is between 0.34 and 2.98 the null hypothesis of equality of population variances must he accepted or judgment reserved.

2. **Questions**

(a) Since $F_{0.99}(9, 9) = 1/5.35 = 0.187 \leq 5 \leq 5.35 = F_{0.01}(9, 9)$ we see that H_0 would not be rejected with $F = 5$, although the variance of the numerator is 5 times that of the denominator.

(b) Since $F_{0.99}(4, 4) = 1/16 = 0.0625 \leq 16$ minus a tidbit $\leq 16 = F_{0.01}(4, 4)$ we see that H_0 would not be rejected with an F almost equal to 16, although the variance of the numerator is almost 16 times that of the denominator.

(c) (i) Since $F_{0.95}(6, 6) = 1/5.05 = 0.198 \leq 5 \leq 5.05 = F_{0.05}(6, 6)$ we see that H_0 would not be rejected with an $F = 5$ at the 0.05 level of significance with sample sizes of 6, although the variance of the numerator is 5 times that of the denominator.

(ii) Since $F_{0.95}(3, 3) = 1/19 = 0.053 \leq 16 \leq 19 = F_{0.05}(3, 3)$ we see that H_0 would not be rejected with an $F = 16$ at the 0.05 level of significance, although the variance of the numerator is 16 times that of the denominator.

(d) This suggests that when testing for equality of variances with small sample sizes we are unlikely to reject H_0 unless one of the variances is much greater than the other. This implies that we are dealing in these cases with a high value of β. That is, there will be many instances where the variances are not equal but, because of the high multiple required for rejection, we wind up accepting H_0 and making a Type II error.

2

If the Strings Attached are Detached in an Application, What Then? Robustness to the Rescue?

2.1 PREFACE

In the sample of recently published basic statistics books that I examined attention given to robustness ranged from shortshrift to not mentioned.

2.2 ROBUSTNESS

A theorem states that if a population is normally distributed, then the sampling distribution of the mean is normally distributed for all sample sizes n.

Suppose the population is "close to being normally distributed", as in the case with applications, what can be said about the sampling distribution of the mean? Assuming that the population has finite variance, we can conclude that for sufficiently large n the sampling distribution of the mean is approximately normal.[1] Since with a "small deviation" from the strings attached to the theorem, its conclusion remains intact that the sampling distribution of the mean is approximately normal, we say that this theorem is **robust** with respect to the normality string attached.

This basic idea of robustness with respect to a string attached is carried over to other situations as well.

---

[1] This situation, as my colleague Michael Kazlow pointed out to me, provides us with another example of the wisdom of taking strings attached seriously. If the population does not have finite variance and we have a sequence of Cauchy distributions, they do not converge to a normal distribution as n $\to \infty$.

As observed in Sec. 1.1, the hypothesis test

$$H_o : \sigma^2 = \sigma_o$$

is **not** robust with respect to the population normality string attached.

In math modeling taken up in Ch. 12 we consider Carl Cairns' problem which requires consideration of corn yield populations obtained by employing the currently used fertilizer and a new one. A question of interest is: are these corn yield populations normal /robust?

Support for or refutation of this condition can be obtained by employing the nonparametric Lilliefors test.

2.3 THE LILLIEFORS TEST

The randomly chosen sample of corn yields for the currently used fertilizer (in bushels per acre) are; 70.5, 80.2, 90.3, 60.2, and 85.4. It is necessary to write them in increasing order: 60.2, 70.5, 80.2, 85.4, 90.3.

The null and alternative hypotheses are:

H_o: The underlying population of corn yields for the currently used fertilizer is normal.

H_a: The underlying population of corn yields for the currently used fertilizer is not normal.

Take $\alpha = 0.05$

From the afore ordered sample we calculate S(x), the sample cumulative distribution function (Table 2.1).

Table 2.1

x	60.2	70.5	80.2	85.4	90.3
$S(x)$	1/5	2/5	3/5	4/5	1

From the sample we compute $\bar{x} = 77.3$ and $s = 12.1$. We next calculate the hypothetical cumulative distribution $F(x)$ under H_0. For example:

$$F(60.2) = P(X \leq 60.2)$$
$$= P\left(z \leq \frac{60.2 - 77.3}{12.1}\right)$$
$$= P(z \leq -1.41)$$
$$= 0.0793$$

The cumulative distribution function $F(x)$ values are shown in Table 2.2.

Table 2.2

x	60.2	70.5	80.2	85.4	90.3
$F(x)$	0.0793	0.2877	0.5948	0.7486	0.8577

We next calculate $|F(x) - S(x)|$ for the sample values and determine the largest such value D, as shown in Table 2.3

Table 2.3

x	$F(x)$	$S(x)$	$\lvert F(x) - S(x)\rvert$
60.2	0.0793	0.2000	0.1207
70.5	0.2877	0.4000	0.1123
80.2	0.5948	0.6000	0.0052
85.4	0.7486	0.8000	0.0514
90.3	0.8577	1	0.1423 ← D

If the null hypothesis is true, then $S(x)$ and $F(x)$ should be close for all x. Large differences between $F(x)$ and $S(x)$ for at least one value of x would be evidence against the null hypothesis.

The Lilliefors test makes this precise by defining the test statistic D as the largest of the differences between $F(x)$ and $S(x)$ in absolute value.

$$D = \text{Largest of } |F(x) - S(x)|$$

The decision bounds for the Lilliefors test in terms of sample size and significance level are given in Table E, part of which is reproduced as Table 2.4.

Table 2.4

Decision Bounds for the Lilliefors Test

Sample size n	Significance Level α		
	0.10	0.05	0.01
4	.352	.381	.417
5	.315	.337	.405
6	.294	.319	.364
7	.276	.300	.348
8	.261	.285	.331

For $n = 5$ and $\alpha = 0.05$ the decision bound is 0.337. $D = 0.1423 < 0.337$ and thus we accept the null hypothesis that the population of corn yields for the currently used fertilizer is normal, or reserve judgment.

Since this test of normality is one part of a preliminary analysis on the populations of corn yields obtained by use of the two fertilizers, we accept the null hypothesis (keeping in mind the possibility of a Type II error).

What the test actually establishes is that the population of corn yields obtained by use of the current fertilizer is close enough to normality (robust with respect to this condition) that it passes the test for normality.

2.4 FOOD FOR THOUGHT

1. **Normality /Robustness of Corn Yields.**

   Test the assumption of normality for the population of corn yields obtained from the new brand of fertilizer at the 0.05 level. The sample yields (in bushels per acre) taken from 5 randomly chosen plots so treated are 73.2, 82.1, 93.1, 63.6 and 87.1.

2. **Normality /Robustness of the Holding Times of *Basic Statistics***

   A recent edition of *Basic Statistics* was the focus of a wrist strength test at Ecap University. The time lengths, in minutes (to one place), that the population of 156 students at Ecap U. who used the text were able to hold it in one sitting is given in Table 2.5. The following random sample was obtained: 3.2, 5.0, 7.2, 4.7, 2.5.

Table 2.5

	Column												
Row	1	2	3	4	5	6	7	8	9	10	11	12	13
1	1.7	4.3	6.0	4.7	1.7	7.2	3.1	3.9	3.3	7.0	2.5	4.9	5.2
2	2.3	4.5	6.9	3.4	3.8	6.6	3.6	3.5	5.5	5.4	5.9	6.1	4.4
3	7.9	4.7	5.1	6.0	5.9	8.4	5.1	6.3	5.2	4.4	5.2	3.0	4.4
4	4.3	3.3	5.0	3.8	7.0	2.9	3.0	4.1	4.5	5.2	3.3	4.2	2.5
5	5.1	2.1	4.6	2.9	4.1	5.9	4.7	4.8	4.1	5.5	5.0	2.6	4.4
6	6.1	5.0	6.3	4.6	8.1	4.6	6.1	6.6	3.2	5.9	1.8	3.2	1.5
7	5.7	7.3	4.3	3.2	3.4	4.1	5.0	1.7	4.3	4.2	3.5	4.1	4.5
8	4.8	3.7	3.0	4.9	7.2	4.5	4.9	6.5	3.1	5.2	4.8	3.9	4.9
9	7.5	3.8	4.5	6.8	4.1	0.3	5.9	2.4	3.6	3.4	1.9	3.4	1.6
10	3.5	3.4	3.6	2.6	5.9	5.1	4.9	6.2	5.0	3.9	6.1	5.6	4.6
11	7.5	6.7	4.4	5.8	4.5	5.1	5.3	4.4	5.8	4.7	2.5	3.5	7.6
12	4.3	4.7	5.1	5.8	3.1	2.3	5.7	8.2	0.8	2.5	7.2	2.6	6.7

Carry out a Lilliefors test on the afore sample to test for the normality of the wrist strength time population. Use the 5% significance level.

2.5 Answers/Discussion of Food for Thought

1. **Normality/Robustness of Corn Yields**

    H_o: The population of corn yields obtained from the new fertilizer is normal.
    H_a: The population of corn yields obtained from the new fertilizer is not normal.
    $\alpha = 0.05$
    The sample cumulative distribution function $S(x)$ is given in Table 2.6:

    Table 2.6

x	63.6	73.2	82.1	87.1	93.1
$S(x)$	1/5	2/5	3/5	4/5	1

    $\bar{x} = 79.8$ and $s = 11.6$, leading to the hypothesized cumulative distribution function $F(x)$ given in Table 2.7.

    Table 2.7

x	63.6	73.2	82.1	87.1	93.1
$F(x)$	0.0823	0.2843	0.5793	0.7357	0.8729

    This yields Table 2.8 and $D = 0.127$.

    Table 2.8

x	$F(x)$	$S(x)$	$\|F(x) - S(x)\|$
63.6	0.0823	0.2000	0.1177
73.2	0.2843	0.4000	0.1157
82.1	0.5793	0.6000	0.0207
87.1	0.7357	0.8000	0.0643
93.1	0.8729	1	0.1271 ← D

The decision bound for this Lilliefors test is 0.337. Since 0.127 < 0.337, we accept/r.j. on H_o.

2. **Nomality/Robustness of the Holding Times of *Basic Statistics***

H_o: The population of wrist strength times is normal;

H_a: The population of wrist strength times is not normal.

$\alpha = 0.05$. $D = 0.197 < 0.337$. Accept/r.j. on H_o.

3

If the Strings Attached are Detached in an Application and Robustness Fails, What Then?

3.1 Preface

If robustness fails, what then? I would look for a nonparametric alternative.

3.2 Mathematics Review Tapes

Ecap University has been giving thought to what could be done to help entering freshmen improve their mathematics level of preparation prior to their taking the required initial year of freshman mathematics. One suggestion being considered would have the students view a series of six one-hour mathematics review videotapes and work through the exercises in a workbook which accompanies the tapes. A faculty member would be available as a resource person. This suggestion met with a positive response, but its implementation on a large scale is formidable and expensive, with questions about its effectiveness.

Before implementing the program on a large scale it was decided to conduct a paired sample study. The admissions department determined a pool of entering freshmen who were willing to participate in the study. Twenty students were chosen at random from the pool and paired on the basis of similar academic profiles. One member of each pair took the videotape review program and the other did not. All took the same initial semester of mathematics with the same instructor. The differences in final exam test scores of the pairs was to be examined at the 0.05 level of significance. The results are given in Table 3.1.

Table 3.1

Pair	1	2	3	4	5	6	7	8	9	10
Took Video Program (x_i)	85	72	66	59	67	91	73	72	76	88
Did not take video program (y_i)	76	64	67	61	65	83	72	74	77	80

(a) State the strings attached on which this hypothesis test is based.

(b) Determine the accept/r.j. and reject regions.

(c) Carry out the hypothesis test.

In answer to (a), the populations of final exam grades of those who take the videotape review program and those who do not take the videotape program must be normal/robust.

Application of the Lilliefors test at the 5% significance level to the afore sample of those who took the videotape review program does not support the normality/robustness of this population.[1]

Since a basic string attached is not attached in this situation it would be meaningless to address (b) and (c). The results obtained would not be credible.

The nonparametric paired-Sample Sign Test comes to our rescue.

3.3 A PAIRED-SAMPLE SIGN TEST

This sign test can be used to analyze data that consist of matched pairs. The null hypothesis for matched pairs is that the population of within-pairs

---

[1] H_o: The final exam grades of those who take the videotape review program are normal/robust . . .

H_a: The final exam grades of those who take the videotape review program are not normal/robust.

$\alpha = 0.05$, $D = .660 > .239$; reject H_o and accept H_a.

differences has a median equal to zero, versus some appropriate one or two sided alternative hypothesis.

With the nonparametric paired sample approach we replace each pair of data values with a plus sign if the first component exceeds the first. No ties are allowed, so that those pairs for which the two components are equal are discarded from the sample. We may then proceed with the z-statistic.

The paired-sample sign test is applicable to populations whose data scale is ordinal or higher.

3.4 Return to the Mathematics Review Tape Situation

For the comparison to be meaningful we must assume (string attached) that the population of paired differences is symmetric so that its mean and median are equal.

Does the mathematics review program help increase students' median exam performance? To start, we reproduce Table 3.1, which lists the final exam scores of the ten randomly selected students who participated in the study.

Table 3.1

Pair	1	2	3	4	5	6	7	8	9	10
Took video program	85	72	66	59	67	91	73	72	76	88
No video program	76	64	67	61	65	83	72	74	77	80

If the video review program helps increase the students' median exam performance, we would expect the median exam grade of the students who participated in the program to "significantly" exceed that of their counterparts who did not participate in the program. That is, most of the test score differences (video program minus no video program) should be positive.

Take:

$$H_o: \pi = 0.5$$

$$H_a: \pi > 0.5$$

$$\alpha = 0.05,$$

which yields Figure 3.1.

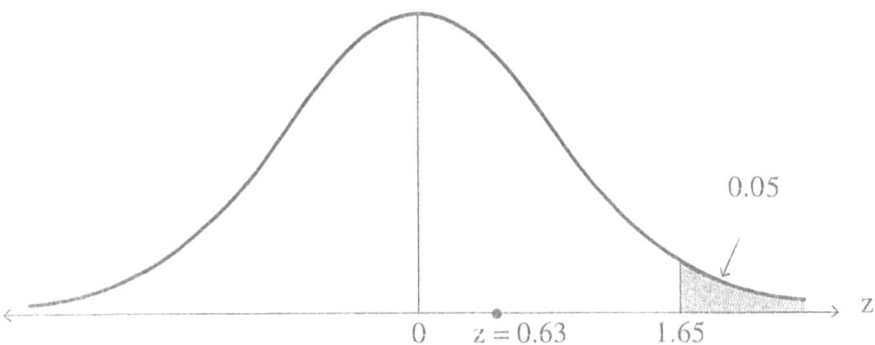

Figure 3.1

The final exam scores of students who did not participate in the video review program exceeded that of their counterparts in pairs 3, 4, 8 and 9, while in the other six pairs, the student participating in the program achieved a higher final exam score. We have

$$+ + - - + + + - - +,$$

six of which are plus, so that $p = 6/10 = 0.6$. The z-statistic may be used since $n\pi = n(1-\pi) = 5 \geq 5$.

$$z = \frac{0.6 - 0.5}{\sqrt{\frac{(0.5)(0.5)}{10}}} = 0.63 < 1.65$$

At the 0.05 significance level we cannot reject H_o. Either we conclude that the videotape review program is not of benefit in preparing freshmen students for their initial study of mathematics, or reserve judgment.

3.5 FOOD FOR THOUGHT

1. **The Mathematics Review Tape Situation**

   In connection with the mathematics review situation described in the afore, it was suggested that the statistical study of the effectiveness of the videotapes be conducted as follows: Choose 6 students at random from those who agreed to participate in the study, have them take the videotape review program, and place them into the first semester mathematics course. Independently, choose another 6 students at random from those who agreed to participate in the study and place them in the same course with the same instructor without having been exposed to the videotape review program.

   (a) If feasible, conduct a hypothesis test on the null hypothesis $H_o: \mu_1 - \mu_2 = 0$ versus $H_a: \mu_1 - \mu_2 > 0$, where μ_1 and μ_2 are the mean final exam scores of the freshmen populations who had and did not have the videotape review, respectively.

   (b) If it is not feasible to conduct this hypothesis test, explain why.

2. **The Effect of Temperature on Skiing Time**

   To study the effect of temperature on skiing time fourteen randomly selected skiers were clocked on a course (in minutes) on two days, once when the temperature was 20°F and once when it was just above the freezing mark. Wind and cloud conditions were similar on the two days. The skiing times are shown in Table 3.2.

   Table 3.2

Skiers	Esther	Joshua	May	Bruce	Carol	Bob	Darice	Barry	Helen	Lou	Sharon	Heather	Rick	Donna
20°F	8.1	7.9	10	8.6	7.8	9.2	8.6	8.4	8.5	9.1	8.5	8.9	7.9	8.8
35°F	8.4	8.2	9.7	8.8	7.9	9.7	8.2	8.3	8.6	9.1	8.4	9.2	8.1	9.2

The feeling was that the increase in temperature might result in an increase in the median skiing time.

(a) Test, at the 0.05 level of significance, the hypothesis that temperature has no effect on the median skiing time against the alternative hypothesis that the warmer temperature adversely affects it.

(b) Conclusion? How so?

3. **The Arnold Clinic's Diet Plan**

The Arnold Clinic wants to test the effectiveness of a proposed diet plan. Does the plan help to reduce the weight of those who undertake it? Weight reduction data for a random sample of 15 dieters who completed the plan are given in Table 3.3.

Table 3.3

Weight Before	137	156	163	174	160	207	184	245
Weight After	142	147	158	180	160	196	173	231

Weight Before	185	172	133	253	129	175	197
Weight After	178	174	138	237	126	164	191

(a) Test the null hypothesis that the diet is not effective in reducing a person's median weight against the alternative hypothesis that it is effective in reducing the median weight. Use the 0.025 significance level.

(b) Conclusion? How so?

3.6 Answers / Discussion of Food for Thought

1. **Mathematics Review Tapes**

   (b) As noted, application of the Lilliefors test to the sample of those who took the video tape review program does not support the normality / robustness of the underlying population:

   It is thus not feasible to conduct the hypothesis test.

2. **The Effect of Temperature on Skiing Time**

   (a) If the warmer weather does "adversely" affect skiing performance time, then we would expect the time in the warmer temperature to be the higher of the two times for most skiers, so that it is reasonable to expect most of the differences (colder temp.—warmer temp.) to be negative. Then there should be relatively few positive signs, which leads us to formulate our null and alternative hypotheses as follows:

   $$H_0 : \pi = 0.5$$

   $$H_a : \pi < 0.5$$

May, Darice, Barry and Sharon took longer in the lower temperature than in the higher temperature, Lou's times in the two temperatures were the same, while all other skiers took less time in the lower temperature. As a result, the sequence of signs is:

$$- - + - - - + + - + - - -$$

Our sequence of 13 signs contains 4 plus signs, so that $p = 4/13 = 0.308$. The z-statistic may be used since $n\pi = n(1 - \pi) = 13(0.5) = 6.5 \geq 5$. $\alpha = 0.05$, gives us Figure 3.2

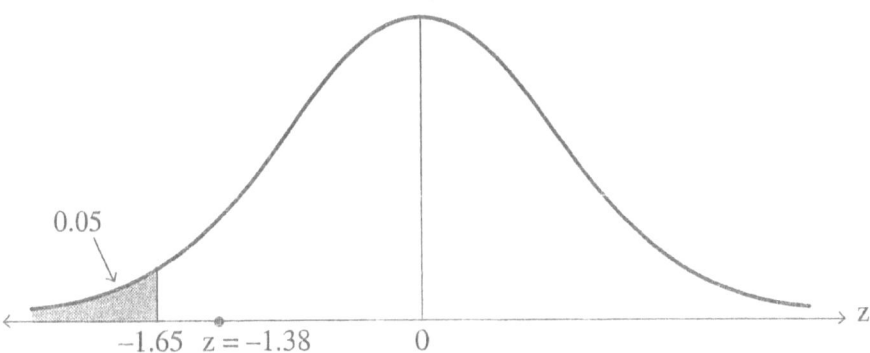

Figure 3.2

$$z = \frac{0.308 - 0.5}{\sqrt{\frac{(0.5)(0.5)}{13}}} = -1.38 > -1.65$$

At the 0.05 significance level we cannot reject H_o. Instead, we accept H_o that the warmer temperature does not have an adverse affect on skiing time or reserve judgment.

3. **The Arnold Clinic's Diet Plan**

   $H_o : \pi = 0.5$; $H_a : \pi > 0.5$; $\alpha = 0.025$.

   The sequence of plus and minus signs is: − + + − + + + + − − + + + +.

   $z = 1.60 < 1.96$. Accept/r.j. on H_o.

4

Good Data, Good Data, My Kingdom for Good Data (Part 1): Is Random Sampling in Practice as Simple as It Sounds in Theory?

4.1 Preface

The idea of random sampling is simple; there is to be no bias, deliberate or inadvertent that favors certain samples of the same size being chosen over others—a level playing field, so-to-speak. We read in statistics books statements such as, choose a sample at random, we assume that a sample is chosen at random, and the like.

What about achieving random sampling

ALARMING! THE CHASM SEPARATING
BASIC STATISTICS EDUCATION FROM ITS NECESSITIES

in practice. Not a word about this problem in the sample of basic statistics books I examined, which in my view is a glaring omission.

Yet, what could be said to enhance our students' perspective on the problem. Consider:

4.2 CAN WE TRUST TV RATINGS?

The life span of a television program is determined by the public's reaction to it, which is measured by TV ratings. These ratings, produced by the Nielsen Company, estimate the audience in terms of the percentage of those sets in use which are turned to each channel, called a share, or in terms of the percentage of the total possible audience, sets on or off, called a rating. Shares and ratings are further broken down according to the sex and age of viewers so that advertisers can better focus their advertising campaigns. These numbers determine the buying and selling of billions of dollars of television air time. They mean life or death to television programs.

The half-hour comedy *Good & Evil*, which had promising ingredients in terms of writing, acting and production talent, had a short life after its premiere in the Fall of 1991 because of low initial ratings. In March 1992 NBC announced that they were dropping two successful shows, *Matlock* and *In the Heat of the Night*, because the demographic numbers favored older viewers while the network wished to build around a more youthful audience.

Since 1986 the data which underlie the ratings have been collected by a device called a people-meter. The remote control part of a people-meter rests on top of the television set. When the set is turned on, the meter prompts viewers to enter their identification number. Information is provided on what channels are being beamed into the household and who is watching them. Nielsen puts its people-meter into 4000 households selected at random from the approximately 93 million homes in America with television.

The people-meter data gathering system produced lower ratings for the networks than had been expected and a serious question arose as to whether this was because of the increased or decreased accuracy of this system over the method it replaced. The networks commissioned a study of the Nielsen methodology and two years later this Committee on Nationwide Television Audience Measurement (CONTAM) issued a nine-volume report that was highly critical of the Nielsen system. The report found evidence of button fatigue—that over time people did not push the buttons that would insure data accuracy as they did in the beginning. CONTAM was highly critical of Nielsen's sampling procedures for obtaining the 4,000 households that make up their sample; random sampling was envisioned in the methodology, but the actual sampling deviated significantly from this requirement. From this came ratings which were highly suspect.

4.3 Going to War

In 1969 the administration of the draft in the United States to determine the order in which men born in 1950 would be drafted was changed to a lottery system. Three hundred sixty six capsules were prepared (for a leap year), each containing a birthdate. Each month's capsules were put into a separate box. The boxes were emptied into a drum, first those for

January, followed by those for February, and so on for the subsequent months. The drum was rotated a few times, the capsules were poured into a bowl, and on December 1, 1969 the drawing was made. Those with birthdays on the capsules drawn first would be drafted first, and so on. If your birthday fell among those drawn last, there was a good chance that you would not be drafted at all. The results of the drawing are given in Table 3.1 from which we see that the earlier months, January through June got the larger share of the last-to-be drafted numbers and the later months, July through December, got the larger share of the first-to-be drafted numbers.

Table 3.1

Month	1—122 (First-Drafted)	123—244 (Middle)	245—366 (Last Drafted)
January	9	12	10
February	7	12	10
March	5	10	16
April	8	8	14
May	9	7	15
June	11	7	12
July	12	7	12
August	13	7	11
September	10	15	5
October	9	15	7
November	12	12	6
December	17	10	4

The results suggest the possibility that the earlier months' capsules were concentrated at the bottom of the bowl, while those of the later months were concentrated at the top and were more accessible for picking. Formal hypothesis tests of randomness did not support the hypothesis that a random drawing had been carried out (see [2] and [3]).

4.4 Randomness Achieved in Practice? Hypothesis Tests Say No, Reality Agrees.

The afore provides one example. Another is provided by the study of the behavior of atoms in a magnetic crystal.

The study of many complex phenomena requires the generation of large streams of random numbers. It came as quite a shock when three scientists showed that five of the most often used computer programs for generating random numbers induced errors in the study of the behavior of atoms in a magnetic crystal because the numbers produced were not random, despite the fact that they passed several statistical tests for randomness [1]. The deviations from randomness were subtle and, although the pseudorandom numbers produced were satisfactory for many purposes, they were not satisfactory for the problem at hand. Is it possible that no machine based system can produce truly random numbers?

John von Neumann thought that the answer is yes. In an observation made in 1951 von Neumann expressed the view that anyone who believed a computer could produce truly random numbers was living is a state of sin. It may be that the best we can hope to do is produce random/robust numbers which are satisfactory for the purpose at hand, and that the truly random number is a mathematical ideal which cannot be attained. The question that arises in an application is, how random/robust is good enough?

References

1. M. Browne, "Coin-Tossing Computers Found to Show Subtle Bias," *The New York Times,* Jan. 12, 1993, p. C1.

2. C. Hawkins, J. Weber, *Statistical Analysis: Applications to Business and Economics* (Harper and Row, 1980), 297-303.

3. J. Rosenblatt, J. Filliben, "Randomization and the Draft Lottery," *Science,* 171 (1971), 306-308.

4.5 FOOD FOR THOUGHT

1. To carry out a marketing survey on consumer preferences for kitchen appliances Elias Marketing Research Associates placed two interviewers on the busiest street in town to interview passersby. Does the sample of opinions obtained qualify as a random sample?

2. Tickets were sold at Ecap University's graduation celebration to help raise funds for the University's new library. The tickets sold were placed in a bowl as soon as they were sold. At the end of the graduation festivities the University's Library Director, Harriet Warren, reached into the bowl and chose a ticket at random. The ticket holder was awarded a newly published edition of Charles Dickens's collected works. Kevin Reynolds, who was among the first to purchase a ticket, protested that the drawing procedure was

biased and demanded that ticket purchasers be given a refund or that the drawing be held again. "Your claim is not justified Mr. Reynolds," replied the Dean of Student Affairs. "Ms. Warren was blindfolded and the ticket was chosen at random." Who is right?

4.6 Answers/Discussion of Food for Thought

1. No. The make-up of the busiest street is not necessarily (and probably not) the same as the make-up of all streets in the city.

2. On the face of it the procedure seems reasonable enough for the task at hand, but how close does it come to satisfying the requirements of a random drawing? The problem is with the physical stirring of the tickets to achieve a "thorough mix." Obtaining a "thorough mix" becomes more and more difficult to achieve as the number of tickets increases, and it is not clear whether early, middle, or late ticket buyers might be favored and to what extent.

   If the tickets were "thoroughly" mixed, the Dean is right. If not, Reynolds is right.

5

Good Data, Good Data, My Kingdom for Good Data (Part 2): How Reliable are These Data? How Relevant are They to the Situation Under Study?

5.1 Preface

The basic statistics texts I am aware of say a good deal about the methods of descriptive statistics, but nothing about the quality of the data they are being applied to. Are the data reliable? Are they relevant to the study being undertaken? Important questions for a statistics course that is application oriented.

My experience is that this neglect leads students to the view that "good data" is a matter to be taken for granted and that the methods of descriptive statistics will yield wisdom, no matter what. GIGO (Garbage in, Garbage Out) has not yet taken hold with full appreciation.

I invite you to consider the following examples which I have used in teaching statistics. More recent versions of these situations are readily available.

5.2 ARE THE DATA RELIABLE?

Case 1. Are Statistically Dangerous Schools Necessarily Dangerous? Are Statistically Safe School Necessarily Safe?

In July 1986 the New York City Board of Education issued a list of its most dangerous schools based on incident and crime reports that it had received. One respondent notes that he never felt unsafe or threatened in teaching at the fifth listed "most dangerous" junior high school. [Spector; 20] Another respondent comments: "I believe that the scorecard . . . names not the most dangerous schools but the schools whose administrators have the courage to report what is really happening." [Richman; 17]

In June 1994 New York City Schools Chancellor Ramon Cortines rejected data on violence in the city's schools, saying that he suspected school administrators were underreporting acts of violence to make their schools appear less turbulent. [Dillon; 6].

In September 1995 Edward Costikyan, the chairman of the Mayor's Commission on School Safety, observed: 'There seems to be a total absence of any reliable numbers on anything. How can you manage anything without knowing what you're dealing with.' [Toy; 25]. Are things better now? . . .

In September 2007 New York City comptroller William C. Thompson Jr. stated that an audit showed that the City had not ensured that all principals accurately report violence in their schools, making it difficult for the public to assess their safety. [Gootman; 10]. Are things better now?

Case 2. Are American Students Really That Bad in Math and Science?

Every few years another study appears which shows again that American students are worse in math and science than their counterparts in even the poorest countries. But are they being compared with their counterparts? Some say no. Without denying that there is much to be improved in American education, a growing number of critics have argued that the test results are flawed because American students in total are consistently being compared with the elite students of other countries. [Kolata; 13]

Case 3. These Data May Give You Nightmares

Halcion, manufactured by the Upjohn Company and introduced in the United States in 1983 is one of the world's best known sleeping pills. Its main advantage over competing products, Upjohn has claimed, is in encouraging nighttime sleep without daytime drowsiness.

How safe is Halcion? It received Food and Drug Administration approval and its manufacturer claims that it is just as safe as other drugs of its kind. Dissenters argue that Halcion is more likely to cause symptoms such as amnesia, paranoia, and depression and that Upjohn engaged in data manipulation to conceal its side effects. This view emerged from a law suit filed by Ilo Grundberg, who killed her mother the day before her mother's 83rd birthday and placed a birthday card in her hand. Mrs. Grundberg claimed that Halcion had made her psychotic, and charges against her were eventually dismissed. Upjohn settled the lawsuit with Mrs. Grundberg before it was to go to trial in August 1991, but in preparation for the suit it had to make available a good deal of data about Halcion to the plaintiff's attorneys.

Dr. Ian Oswald, who was head of the department of psychiatry at the University of Edinburgh and spent 30 years doing research on sleep, was obtained as an expert witness. Dr. Oswald spent two years going over Upjohn's data and concluded that Upjohn had known about the extent of the drug's adverse effects for 20 years and concealed these data. He concluded that "the whole thing had been one long fraud." [Kolata; 14]. Dr. Graham Dukes, former medical director of the Dutch drug regulatory agency, who examined some of Upjohn's data, believed that the data on Halcion had been organized in such a way as to minimize the drug's adverse effects and that this could not have occurred accidentally.

In reaction to the criticisms voiced, Britain, the Netherlands and Belgium were led to remove the sleeping pill from the market. A report issued in April 1994 by F.D.A. investigators stated that the Upjohn Company had engaged in ongoing misconduct with Halcion.

Case 4. Are These Nutrition Data Healthy?

In November 1993 the General Accounting Office reported that the nutrition data in the publication Handbook 8 is flawed because of sloppy, inconsistent, or questionable methods of collecting data. Handbook 8 states, for example, that there are 3000 international units of Vitamin A in a portion of papaya while it is known that there are 400 units in the same portion. Nutrition data on bacon-cheeseburgers, to take another example, comes from brochures provided by fast food chains, which do not explain the basis for their claims.

Handbook 8 is a major resource of nutrition data. It plays a role in determining public nutrition policy, planning feeding programs, medical research, and providing information for folks on diets who are trying to keep track of such things as calories, fat levels, and sodium intake.

Case 5. Which Came First: The Numbers or the Decision?

By the early '90s there was general agreement that some military installations should be closed as a cost cutting measure. The question is, which ones? Maybe yours, but not mine, was the answer offered by local politicians as the struggle to save local bases heated up.

The Rome Laboratory in Rome, New York, a high-tech Air Force research installation, came under scrutiny. An analysis carried out in October of 1994 of a proposal to close the Lab and move its facilities to Hanscom Air Force base in Massachusetts led to the conclusion that it would cost $133.8 million to move the Lab with annual savings of $1.5 million being realized. It would take more than 100 years to recover the cost of the move at this rate. The Pentagon did not recommend closing the Rome Lab.

In February of 1995 the Air Force released an analysis of a proposal to close the Rome Lab and move 60 percent of its operations to Hanscom and 10 percent to the Army base in Fort Monmouth, New Jersey. It led to the conclusion that relocating the Lab would cost $52.8 million with annual savings of $11.5 million. The cost of the move would be recovered in four years at this rate. The Pentagon recommended closing the Rome Lab.

In May 1995 another analysis led to the conclusion that relocating the Lab as described would cost $79.2 million with annual savings of $13 million. The cost of the move would be recovered in six years at this rate. Again the Pentagon recommended closure.

Which numbers, if any, are realistic? They arise from different assumptions, which means that the question of number realism falls back on the question of assumption realism. Did politics play a role in the decision to close the Rome base? Senator Alfonse D'Amato of New York believed that the answer is yes. They reached a back room political decision to close the Rome base and generated numbers that would justify that decision, Senator D'Amato and other New York officials believed. Analysts with the Air Force deny this, and argue that the third analysis is the most comprehensive and realistic. In April of 1995 the General Accounting Office criticized the Air Force for providing insufficient documentation for the analyses that prompted decisions to close bases. This takes us back to square one, the nature of the assumptions.

In June of 1995 a Presidential commission reversed the Pentagon's closure recommendation. It gave a 13 year estimate for the time it would take to recoup the cost of relocating the Lab, which it felt was too long to justify its closure. Is 13 years more realistic than the previous estimates of 100+, 4 and 6 years? Again, this takes us back to the assumptions.

Case 6. Assumptions Driven By Ideology: Reality's Revenge

As President Jimmy Carter was preparing to leave office in January 1981, the first priority of the new Reagan Administration was to conduct a thorough overhaul of the Carter budget for fiscal 1982, which was to begin on October 1, 1981. David Stockman was appointed Director of the Office of Management and Budget and Reagan's budget team set to work in January 1981.

Stockman notes: "There were three doctrines represented on the forecasting team: the monetarists, the supply siders, and the eclectics. . . . the new chairman of the Council of Economic Advisors, Murray Weidenbaum, tended toward the third approach." [Stockman; 21]

In conversation with the journalist Laurence Barrett [Barrett; 1], Weidenbaum commented that: "It was a forced marriage. The supply-side people insisted on [forecasting] rapid growth in real terms and the monetarists insisted on rapid progress in bringing down inflation. Each of them would go along with a set of numbers as long as their own concern was satisfied. The monetarists weren't that concerned about growth and the supply-siders weren't that concerned about inflation."

Weidenbaum was a latecomer to the initial negotiations on economic assumptions. Barrett [Barrett; 2] reports that "he was so shocked at what he found that he seriously considered resigning even before unpacking the cartons in his new office. He had the same urge a few months later, when the assumptions were reviewed and retained for political reasons, though by then everyone knew they were specious." Weidenbaum stayed until the summer of 1982 because he felt that, bad as things were, the economic assumptions would have been more irresponsible than they were had he not been there.

Stockman [Stockman; 22] notes: "The table that follows [5.1] tells the whole story, proving that our Rosetta stone was a fake. . . . The February 1981 economic forecast eventually became known as 'Rosy Scenario.' Weidenbaum wrote the final specific numbers. But its underlying architecture—the push-pull hypothesis—was ultimately the work of a small band of ideologues."

Table 5.1

Real GNP Growth (%)

Quarter	Supply-Side/ Monetarist Consensus	Final Weidenbaum Forecast	Actual Outcome
1981:4	4.0%	4.0%	−5.3%
1982:1	9.4%	5.2%	−5.5%
1982:2	7.8%	5.2%	0.9%
1982:3	6.8%	5.2%	−1.0%
1982:4	5.4%	5.2%	−1.3%

Case 7. Top of the Line Deception

In 1992 the General Accounting Office audited seven "Star Wars" tests conducted between 1990 and 1992. It found that four of the test results described to Congress as successes were false whereas the three tests that were described as complete or partial failures were correct. [Weiner; 26]

Case 8. Number Magic

In early 1981 Ronald Reagan's budget director David Stockman found himself drowning in a lake of red ink. President Reagan had promised that the country would have a balanced budget by 1983, and Stockman found himself forced to move the target date to 1984.

> Stockman writes: [Stockman; 23]:
> But that was merely a straw in the wind compared to what would come next. I soon became a veritable incubator of shortcuts, schemes, and devices to overcome the truth now upon us—that the budget gap couldn't be closed except by a dictator.
>
> The more I flopped and staggered around, however, the more they went along. I could have been wearing a sandwich board sign saying: Stop me, I'm dangerous! Even then they might not have done so . . .

> Bookkeeping invention thus began its wondrous works. We invented the 'magic asterisk': If we couldn't find the saving in time—and we couldn't—we would issue an IOU. We would call it 'Future savings to be identified.'

> It was marvelously creative. A magic asterisk item would cost negative $30 billion . . . $40 billion . . . whatever it took to get a balanced budget in 1984 after we toted up all the individual budget cuts we'd actually approved.
>
> The magic asterisk passed presidential and congressional muster.

Case 9. Slippery Statistics: An International Dimension

The Slippery Statistics Society (SSS) has an international clientele. Here are a few examples.

Brazil

In a frank conversation between television interviews that was inadvertently broadcasted across his country, Brazil's finance minister Rubens Ricupero expressed the sentiments of many kindred spirits when he confessed of economic indicators: "I have no scruples, what is good we take advantage of. What is bad, we hide" [Brooke; 3]. Minister Ricupero was immediately dismissed, but was this because of his performance or indiscretion?

Britain

In the past *The Economist* has been critical of the U.K. Central Statistical Office as having 'figures often tasting of fudge.' [Duncan and Gross; 7, p. 66], [Economist; 8, p. 88], [Economist; 9, p. 65].

China

Chinese government statistics have run a gamut of slipperyness. After the Communist Party assumed control in 1949, government statistics were systematically distorted to serve the wishes of the new political establishment. During the period of the Cultural Revolution of the late 1960s and early '70s data-gathering was abandoned as unscientific.

Since the passing of the Cultural Revolution, data-gathering and the publication of state statistics has resumed and other pressures have developed. In May 1994 Zhang Sai, director of the State Statistical Bureau, 'warned that distorted statistics are increasing tensions between Beijing and localities.' [Tefft; 24]

Foreign investors in China are wary of Chinese statistics and many have taken to generating their own.

International Atomic Energy Agency

Deaths Untolled

In his engrossing article on the long-term effects of the Chernobyl disaster ["Life in the Zone," Letter from Ukraine, June, 2011 *Harper's Magazine*], Steve Featherstone fails to mention Alexey Yablokov, who recently compiled the findings of some five thousand scientific papers, mostly in the Slavic languages, and determined that in the quarter century since the disaster one million people have died from exposure to radiation.

Yablokov, a respected biologist and member of the Russian Academy of Sciences, visited Washington, D.C., in March and expressed dismay at the silence with which his research was met in the West. Ignoring his report means that we are left with the absurdly low estimate of four thousand fatalities since 1986—a figure provided by the International Atomic Energy Agency, an organization that has long been riven by the conflict of interest between the twin duties of promoting nuclear power and monitoring its safety.

Ralph Nader [Nader; 15]
Washington, D.C.

Japan

By late July 1998 American financial experts reached the conclusion that the magnitude of Japan's banking crisis was far worse than had been publicly acknowledged. The bad debts were estimated as being on the order of $1 trillion, nearly twice the official estimate. The true amount, financial experts emphasized, is hard to pin down because Japanese banks have been using accounting tricks to conceal debts that are not being paid. [Sanger; 18]

Soviet Union and Russia

From the beginnings of the Soviet State, Soviet statistics have acquired a reputation of being unreliable. (See [Clark; 5], [Shaffer; 19].) Writing in 1990, V.N. Kirichenko, Chairman of the USSR State Committee on Statistics, expressed a hope to 'ensure the accuracy of the data . . . restore the trust in such data on the part of the Soviet and international public. The country can no longer afford to seek the right way with the help of trick mirrors.' [Duncan and Gross; 7, p. 66] and [Kirichenko; 12, pp. 50 -57].

Since the breakup of the Soviet Union, Russia has continued to have problems with government statistics, but for different reason. Rather than exaggerating output the statistical pendulum has swung to the extreme of underestimating it. In June 1998 Russia's top statisticians were arrested on charges of manipulating data to underestimate the production of Russian businesses to help them minimize their tax obligations. [Gordon; 11]

United States

In June 1998 the thrust of the Republican majority in the House of Representatives was to cut taxes beyond what was called for in the earlier balanced-budget agreement. But then there are the spending cuts needed to achieve balance. The Congressional Budget Office did not produce the numbers needed for this to work out, which prompted the Republican leadership to address a letter to the Appropriations subcommittee warning that if the C.B.O. did not begin to produce better numbers, 'we must review [its] structure and funding.' [*The New York Times*; 16, A22]

References

1. L. Barrett, *Gambling with History; Ronald Reagan in the White House* (New York: Double day & Co., 1983), P.140.

2. L. Barrett, *Ibid.*

3. J. Brooke, "In Brazil, Slip of the Tongue Makes Campaign Slip," *The New York Times*, Sept. 5, 1994.

4. A. Bryant, "A Different Gauge for Rating Airlines," *The New York Times,* March 7, 1995.

5. C. Clark, *A Critique of Russian Statistics* (London: MacMillan and Co., 1939).

6. S. Dillon, "Report Finds More Violence in the Schools," *The New York Times,* July 7, 1994.

7. J. Duncan, A. Gross, *Statistics for the 21st Century* (Chicago: Irwin, 1995).

8. *The Economist,* "The Good Statistics Guide," Sept. 7, 1991.

9. *The Economist,* "The Good Statistics Guide," Sept. 11, 1993.

10. E. Gootman, "Undercount of Violence in Schools: Defective Reporting is Found at 10 sites", The *New York Times,* Sept. 20, 2007.

11. M. Gordon, "Moscow Statisticians Accused of Aiding Tax Evasion," *The New York Times,* June 10, 1998.

12. V. Kirichenko, "Return Credibility to Statistics," *Business Economics,* Oct. 1990.

13. G. Kolata, "Which Students are Worst at Science," *The New York Times,* Dec. 24, 1991.

14. G. Kolata, "Maker of Sleeping Pill Hid Data on Side Effects, Researchers Say," *The New York Times,* Jan. 20, 1992.

15. R. Nader, "Deaths Untolled," (Letter), *Harper's Magazine,* Aug. 2011 (p. 4)

16. *The New York Times,* "Rigging the Numbers," June 15, 1998.

17. E. Richman, Letter, *The New York Times,* Aug. 12, 1986.

18. D. Sanger, "Bad Debt Held by Japan's Banks Now Estimated Near $1 Trillion," *The New York Times,* July 30, 1998.

19. H. Shaffer (ed.), *The Soviet Economy: Western and Soviet Views* (New York: Appleton-Century-Crofts, 1963).

20. R. Spector, Letter, *The New York Times,* Aug. 12, 1986.

21. D. Stockman, *The Triumph of Politics: How the Reagan Revolution Failed* (New York: Harper & Row, 1986), p. 92.

22. D. Stockman, *op. cit.,* p. 98.

23. D. Stockman, *op. cit.,* pp. 123-4.

24. S. Tefft, "China Is Under Pressure to Clean Up Its Statistics," *The Christian Science Monitor,* June 9, 1994.

25. V. Toy, "Draft Audit Says Board of Education Underrates Crime in Schools," *The New York Times,* Sept. 2, 1995.

26. T. Weiner, "General Details Altered 'Star Wars' Test," *The New York Times,* Aug. 27, 1993.

5.3 FOOD FOR THOUGHT

The following 15 cases make clear the variety of settings in which questionable numbers find their way to us. In replying to the questions posed consider the articles cited and other relevant information that you might obtain.

1. "These numbers are all extremely fuzzy." How so? E. Schmitt, T. Shanker, "Taliban and Qaeda Death Toll In Mountain Battle Is a Mystery," *The New York Times,* March 14, 2002; Al. B. Bearak, "Afghans Declare Mountain Victory; Foes'Toll Unclear," *The New York Times,* March 13, 2002; Al.

2. Are America's productivity numbers for the late 1990's as good as they once seemed to be? "Measuring the New Economy," "A Spanner in the Productivity Miracle," *The Economist,* Aug. 11, 2001; 12-13, 55-56.

3. Do approved cigarette tests provide reliable measures of tar doses that smokers are subjected to? R. Kerber, "Do Approved Cigarette Tests Understate Tar?" *The Wall Street Journal,* Jan. 30, 1997; B1.

4. Is the sharp drop in crime in recent years real or manufactured? F. Butterfield, "As Crime Falls, Pressure Rises to Alter Data," *The New York Times,* Aug. 3, 1998; A1; A. Fine, "Philadelphia Takes Lead in Cleaning Up Crime Statistics," *The Christian Science Monitor,* Dec. 28, 1998; 2.

5. Numbers pulled from the air? How so? C. Wren, "Phantom Numbers Haunt the War on Drugs," *The New York Times,* April 20, 1997; E4.

6. Russia's economy grew by 5% in 2001, the budget was balanced, and inflation came down. Good numbers, but is there more to it than meets the eye? Scratch and Sniff: Russia's Economic Recovery is More Fragile Than It Seems," *The Economist,* Feb. 16, 2002; 71.

7. What is the problem with African Numbers? N. Onishi, "African Numbers, Problems and Number Problems," *The New York Times*, Aug. 18, 2002; Wk-5.

8. Metropolitan Transportation Authority hanky-panky with millions of dollars to build a case for a transit fare increase? What did the judge say? R. Kennedy, "Hevesi Says M.T.A. Moved Millions to Simulate a Deficit," *The New York Times*, April 23, 2003; B1. R. Kennedy, "Saying Public Was Misled, Judge Rescinds Transit Fare Increases," *The New York Times*, May 15, 2003; A1.

9. Social Security's deficit determined by manipulation on the part of politically appointed trustees? D. Langer, "Cooking Social Security's Deficit," *The Christian Science Monitor*, Jan. 4, 2000; 9.

10. What are "dodgy accounting's greatest hits"? R. Abelson, "Truth or Consequences? Hardly," *The New York Times*, June 23, 1996.

11. What prompts the comment: "In today's financial climate, auditors' reports have about as much credibility as buy recommendations from Wall Street analysts"? D. Henry, "For Accountants, A Major Credibility GAAP," *Business Week*, July 23, 2001; 71.

12. What is "backing in"? A. Berenson, "Tweaking Numbers to Meet Goals Comes Back to Haunt Executives," *The New York Times*, June 29, 2002; Al.

13. Does Congress employ the same cook-the-books methods that made Enron infamous? G.R. Craddock, "Congress Uses Playbook Similar to Enron's," *The Christian Science Monitor*, Feb. 6, 2002; 2.

14. Contingent convertible bonds blindside shareholders through hidden costs? How so? D. Henry, "The Latest Magic in Corporate Finance," *Business Week,* Sept. 8, 2003; 88-9.

15. A case of bi-partisan book-cooking? D. Francis, "Budget Fudge," *The Christian Science Monitor,* Aug. 25, 1999; 8.

5.4 Answers/Discussion of Food for Thought

Quotes are from the articles listed in Sec. 5.3.

1. After the start of a military operation the Pentagon listed the confirmed number of enemy dead at 517. Another 250 were believed to have been killed. A few days later the total estimate had risen above 800. 'These numbers are all extremely fuzzy,' one senior military official is quoted as saying.

2. "The miracle of the late 1990s was not quite so miraculous." . . . "Statistical revisions show how America's recent productivity boom is less remarkable than was once thought."

3. A study using methods meant to reflect conditions that are more true to life than those used in Federal Trade Commission approved tests show that the answer to the question posed in the article's title is yes.

4. "In Philadelphia, the city has had to withdraw its crime figures from the national system maintained by the Federal Bureau of Investigation for 1996, 1997 and for at least the first half of 1998 because of underreporting and downgrading crimes into less serious incidents and sloppiness."

    "Gil Kerlikowske, the former Police Commissioner of Buffalo, said the pressure on police departments to prove their performance through reduced crime figures, with promotions and pay raises increasingly dependent on good data, "creates a new area for police corruption and ethics," along with the traditional problems of brutality and payoffs."

    "A common thread running through many of the incidences of police officials altering crime statistics has been that police commanders have downgraded felonies like aggravated assault and burglary, which are reported to the F.B.I., to misdemeanors like vandalism that are not reported to the bureau."

5. "America's drug problem seems impossible to grasp without some sense of its size and scope. But elected officials, and their constituents, want concrete evidence of what is essentially a shadowy illegal activity. And when sensibly vague estimates based on the little that is known won't suffice, law enforcement officials oblige them with numbers that one police officer characterized as "P.F.A.," or "pulled from the air.""

    Statistic: Law enforcement authorities interdict only 10 to 20 percent of the drugs entering into this country.

    Background: "There's no way of telling," said a Government official, who asked for anonymity. "One year you might be seizing 50 percent. The next year you might seize 5 percent. It's a matter of your best guess." The 10 percent figure, by one account, came about a decade ago from a law enforcement official who was pressed for a precise number at a Congressional hearing.

Statistic: Marijuana has quietly become one of the largest cash crops in the United States.

Background: Nobody knows how much marijuana is grown in this country because much of it is cultivated indoors or concealed among other crops. Professor Kleiman said he believed that the notion of marijuana as a vast crop came from an exasperated agriculture official filling out a Government questionnaire about his neglected county in northern California.

6. The World Bank identified three major problems: (1) failure to diversify, (2) low productivity growth, (3) difficulty in implementing reforms passed by parliament.

7. "To policy makers, humanitarian workers or journalists working in sub-Saharan Africa, one of the hardest things to find is a reliable number. Lack of money and expertise, the collapse of roads and railways that has cut off huge swaths of the continent, all make compiling solid statistics nearly impossible. In many countries, very little is known, statistically speaking, outside the capitals. The latest statistics, or the only ones, are sometimes decades old, from colonial days.

    "SOMETIMES, however, figures are based not on scientific estimates, but on pure guesswork. In Africa, in the absence of any figures at all, imaginary ones take on a life of their own—as they did last year with the charges that child workers were forced to work in Ivory Coast's cocoa plantations.

    > Many accounts in the British and American news media last year spoke breathlessly of 15,000 child slaves on Ivory Coast's cocoa plantations, producing the chocolate you eat.

    > The number first appeared in Malian newspapers, citing the Unicef office in Mali. But Unicef's Mali office had never researched the issue of forced child laborers in Ivory Coast. The Unicef office in Ivory Coast, which had, concluded that it was impossible to determine the number.

> *Still, repeated often enough, the number was gladly accepted by some private organizations, globalization opponents seeking a fight with Nestlé and Hershey, and some journalists."*

8. "A state audit concluded" that the Metropolitan Transportation Authority secretly shifted hundreds of millions of dollars in surplus money to create the appearance of a huge deficit in 2003, misleading transit riders about the need for a fare increase this year."

9. "I was surprised to learn that, contrary to the impression these actuaries give that they dictate the crucial economic and demographic assumptions underlying their financial projections, the choice is actually made by the *politically appointed* trustees.

   Briefly, the actuaries provide the trustees with a preliminary report containing financial projections on the basis of recommended assumptions about the future (wage increases, gross domestic product, mortality rates, etc.), along with the projected deficit or surplus, for periods of up to 75 years. . . .

   > *The trustees would then ask questions such as, "What if we lower the fertility rate by 5 percent?" And the senior actuary would then quickly estimate, based on rules of thumb, a deficit of, say, 2.2 percent. At some point, the trustees tell the actuaries the deficit level they desire. The actuaries will then put together the appropriate assumptions and computations for the trustees' annual report.*

   > *This procedure is, of course, improper, because it opens the system to political manipulation by the trustees. They are able, for instance, to make Social Security look as if it is in serious financial trouble and requires radical change to save it.*

10. Recording revenue before it is earned; inventing fictitious revenue; bolstering income with one-time gains; shifting current expenses into a future period; failing to record or disclose liabilities; using hidden reserves to smooth out income; accelerating expenses in a special restructuring charge.

11. "Last year [2000] there were 156 restatements to correct corporate earnings reports . . . They cost investors an estimated $31.2 billion in market value . . . where were the auditors?"

12. "On Wall Street, it is called backing in. Each quarter, analysts at securities firms forecast the profit per share of the companies they cover. Companies whose profit falls short of the consensus estimate can be severely punished, their stocks falling 10 percent or more in a day.

    So some companies do whatever they have to to make sure they do not miss that estimate. Instead of first figuring out their sales and subtracting expenses to calculate the profit, they work backward. They start with the profit that investors are expecting and manipulate their sales and expenses to make sure the numbers come out right.

    During the last decade's boom, as executive pay was increasingly based on how the company's stock performed, backing in became both more widespread and more aggressive. Just how much so is only now becoming clear.

13. Everyone in Congress except the pages—committed to silence on all things political—appears eager to blast Enron for accounting irregularities that obscured the real financial condition of that company.

    But as lawmakers start deliberations on next year's budget, they're cribbing from many of the same recipes that Enron used to cook the books: hiding obligations and overstating revenues.

    It's no secret, and it's not new. For decades, critics on and off Capitol Hill have complained about accounting devices that ignore long-term liabilities, such as Social Security and Medicare, while inflating current revenues.

14. "Pressure to hide profit, not losses."

15. "The idea of 'saving' Social Security with upcoming budget 'surpluses' is sort of funny, considering that the 'surpluses' come entirely from Social Security trust-fund money in the first place."

5.5 ARE THE DATA RELEVANT TO THE SITUATION UNDER CONSIDERATION?

Case 1. Temperature vs. Wind Chill Factor

Richard Browne was informed by Metro Weather that the temperature was 60°. He put on his jacket and stepped outside, intending to take a walk, but within two minutes he was back inside, shivering. Nobody said anything about that wicked wind, he thought, and, turning on the Weather Channel, learned that the wind chill factor was 28°.

Case 2. Which Data "Best" Reflect Airline Reliability?

The long time standard measure of an airline's reliability is its percentage of on-time arrivals, where a flight is deemed on-time if it arrives within 15 minutes of its scheduled arrival time. Such data are widely trumpeted by airlines in their advertising campaigns.

But is this statistic the "best" measure of an airline's reliability? According to Julius Maldutis, an airline analyst with Salomon Brothers, the answer is no. Maldutis argues that a much better measure of reliability is the percentage of flight-miles completed. Look at the cancellation rate, which is indicative of a more troublesome situation to travelers than that indicated by the artificial on-time statistic, says Maldutis. [Bryant; 1]

Case 3. According to the Numbers, the Recession is Over

In early 1992 President Bush was, as modern parlance would put it, an unhappy camper. Government statistics showed a mild recession and strong economic fundamentals. Yet business and consumer confidence in the economy had been shaken to an extent that seemed way out of proportion to the statistical signs, and many were blaming George Bush for having missed the wake up call.

"The problem," observes Charles McMillion [McMillion; 4] "is that many of those statistics are wildly misleading."

One statistic concerns unemployment. During the 1982 recession, the worst since World War II, unemployment reached 10.8 percent. During the 1991-92 recession, it reached 7.8 percent. McMillion notes that the 1991-92 unemployment number looks good by comparison because it mixes two factors, jobs and the size of the labor

force, and neglects the fact that the labor force has contracted sharply. "A better gauge," he argues, "is the number of actual jobs." Three hundred thousand more jobs were lost in the 1991-92 recession than the 1982 recession, June 1981-January 1983. A larger portion of the jobs lost this time involved higher-wage white collar workers, with ramifications throughout the economy. People working or seeking employment has declined by 1.2 million people in the first 19 months of this recession as opposed to 125,000 in the first 19 months of the 1982 recession. These features are not revealed by the unemployment rate.

Manufacturing output is another statistical indicator of economic health. According to this statistic, using constant output values, manufacturing has remained near 22 percent of America's gross national product since World War II. But, McMillion notes: "Even Commerce Department officials who assigned these values admitted—in the Survey of Current Business last year—that "only a substantial research effort over many years holds any promise of overcoming . . . formidable statistical problems with these figures." The rapid pace of technological change makes it virtually impossible to measure "constant" output over time.

U.S. Competitiveness: A comparison of the gross domestic product per worker of the United States against that of other major industrial competitors shows the United States to be well ahead of such rivals as Germany and Japan. "The tally depends," McMillion observes, "on the value assigned to the dollar. . . . Most comparisons use theoretical—so called 'domestic purchasing power parity'—values that vastly over-value the dollar."

The **assumptions** underlying such statistical economic indicators as unemployment, manufacturing output and U.S. competitiveness must be watched. What must also be watched are the limitations of such indicators, and what they omit which is relevant.

Case 4. Can Andy Afford a $200,000 Porsche?

For many of us $200,000 is a considerable sum, and if Andy's financial state were anything like ours, we would probably be inclined to say no. There's a big "if" here, that points to the question which is at the heart of the matter: What is Andy's financial state? Since we don't know, we can only proceed by making some **assumptions**.

Scenario 1.

Andy's after tax assets from salary and investments are approximately $50,000 per year. With a $200,000 obligation against $50,000 in assets per year, it's difficult to see how Andy would be able to avoid defaulting down the road.

Scenario 2.

Andy's after tax assets from salary and investments are approximately $2 million per year. With a $200,000 obligation against $2 million in assets per year, we would probably be inclined to tell Andy to go for it.

The lesson to be learned from Andy's situation is that when it comes to a debt to be carried, focusing on the amount of the debt in absolute terms ($200,000 or whatever) is not the appropriate number trail. The appropriate number trail consists of the ratio of debt to ability to pay expressed by a measure of assets. What about countries? One may ask.

Case 5. The National Debt: Public Nuisance or Menace?

By the beginning of 1993 the gross national debt of the United States, it was generally agreed, was in the neighborhood of $4.2 trillion, a staggering figure which boggles the mind. If you had to transfer this amount of money in $100 bills from one location to another, you would have to deal with a stack of bills 2670 miles long.

The figure sounds ominous, but here is where disagreement begins. One point of view argues that the figure itself and the rate at which it has been increasing portend catastrophic consequences in the offing. When the debt grows faster than the country's ability to carry it, a breakdown with social, political, and economic upheaval is inevitable, and we are coming dangerously close to this state, this view has it.

But is the total size of the debt the figure we should be giving our first priority? Another view argues that in terms of the state of the economy, we are looking at the wrong figure and that, while debt reduction is desirable, it should not be given top priority and carried out in a "mindless" way since this will severely damage the economy. Its proponents focus on the ratio of publicly held debt to Gross Domestic Product (GDP).

In principle, how different is this situation from the one Andy finds himself in?

Case 6. Is Its Gross Domestic Product the "Best" Measure of a Country's Well-Being?

How well off are a country's people? The simplest answer is GDP, which is designed to measure the value of goods and services produced in the country, but does not address such factors as health, employment, destruction of the environment. On September 14, 2009 a commission appointed by Nicolas Sarkozy, President of France, presented its findings in a 292 page report. [Stiglitz; 5]. Also see [*The Economist;* 2] and [Goodman; 3].

Case 7. A Visit to Huxley College

Huxley College, the doings of the Marx brothers (Horse feathers), cannot be located on a map, but its spirit and problems are as real as those colleges and universities which can be so determined.

Ivor M. Wisdom, a financial analyst, was hired by President Marx of Huxley College to analyze the operations of the departments of Huxley College and make recommendations on how to improve their financial efficiency. I.M. Wisdom defined the income of each department as the tuition income of the students being serviced by the department minus costs, primarily salary costs. He collected data on the class size of each instructor and each instructor's rank and salary and found that a number of full professors at the top of the salary scale were teaching classes with a small number of students. To improve the income of the departments, Wisdom recommended that teachers at the top of the salary scale be assigned basic level courses which can be expected to have a large number of students.

There is no question that Wisdom's data were consistent with his view of financial efficiency and reliable, but are they relevant to the issue? "Horsefeathers," cried Dean Harpo Marx, President Marx's brother. "It's irrelevant to the overall financial efficiency of the departments and Huxley as a whole. You could change the teaching assignments at the last minute and this game of academic musical chairs will not change a department's overall tuition revenue or costs on which financial efficiency depends. The scheme is counterproductive in that it deflects us from the real issue of financial efficiency. It may also have negative academic consequences if taken seriously by leading us to assign courses to be taught on the basis of an instructor's spurious personal efficiency rather than academic qualifications."

References

1. A. Bryant, "A Different Gauge for Rating Airlines," *The New York Times,* March 7, 1995.

2. *The Economist*, "Economic Focus: Measuring What Matters," Sept. 19, 2009; 88.

3. P. Goodman, Emphasis On Growth Is Called Misguided, *The New York Times*, Sept 23, 2009, B1.

4. C. McMillion, "Facing the Economy's Grim Reality," *The New York Times*, Feb. 23, 1992.

5. J. Stiglitz (chairman). "Report by the commission on the Measurements of Economic Performance and Solid Progress." Available at www.stiglitz-sen-fitoussi.fr.

5.6 Food for Thought

1. **Decision - Making Framework, RC-1**

   The president of Ecap University charged his Dean of Administrative Affairs, Michael Russell, with the task of setting up a criterion for running or cancelling course sections that would take into account student needs, be sensitive to maintaining academic quality, address the cost dimension, and be simple to use.

   Dean Russell started with the assumption that each section, with perhaps a few exceptions, should pay its own way. He set up a course section run-cancel criterion, RC-1, based on the difference between the tuition revenue generated for the section based on student enrollment and the salary cost of the instructor for the section. RC-1 says run the section if revenue minus cost equals or exceeds $5000.

   $$R - C \geq 5000$$

   Otherwise, it is to be canceled, unless a compelling student need for the course could be established.

   How well does RC-1 satisfy the requirements for a section run/cancel criterion stated by the president of Ecap University?

2. **Decision - Making Framework, RC-2**

Dean Michael Russell of Ecap University was asked to develop other criteria for running or cancelling course sections which takes into account student needs, is sensitive to maintaining academic quality, addresses the cost dimension, and is simple to use. The dean came up with Russell criterion 2, RC-2, which operates as follows: For a given department, English, for example, determine the average salary cost of the faculty in the English Department. Run English section A, let us say, if the tuition revenue based on section A's enrollment equals or exceeds the average salary cost of the English department for section A by $5000. Otherwise, cancel section A unless compelling student or academic needs can be established. The same system would apply to courses in other departments.

(a) What data would be needed for the implementation of RC-2?

(b) What are the merits and disadvantages of RC-2?

5.7 Answers / Discussion of Food for Thought

1. **Decision-Making Framework, RC-1**

    RC-1 gets the job done, but there is a serious question about whether it's the "best" system for handling course run/cancel decision making in a way that satisfies the president's mandate.

    1. **Academic Quality.** Department heads are not free to assign the "best" person to teach a section because the "best" person might be too costly in terms of the $R - C \geq 5000$ condition. A senior faculty member assigned to a course will often carry a much higher cost than a junior or adjunct colleague. If circumstances lead to changes in teaching assignments, courses which would run under one assignment framework might well have to be cancelled under another. With RC-1 the running or cancellation of courses depends more on how the game of academic

musical chairs is played out than on the academic needs for running them, which is an unsatisfactory condition. It is in the overall interest of Ecap U. and, in particular, its students, that department heads be able to assign the most suitable faculty to courses and make changes when circumstances dictate. RC-1 is not compatible with this condition.

2. **Tunnel View.** RC-1 does not take into account the total revenue—total cost picture since reorganization of faculty teaching assignments by itself neither changes the total tuition revenue nor the total cost of faculty salaries.

3. **Data Collection.** The seeming simplicity of RC-1 in terms of data needed for its implementation is deceptive. A good deal of data has to be obtained and evaluated for each section in terms of a particular faculty assignment to the section before a run/cancel decision can be make. This is time consuming, but must be carried out within a tight time frame after class registration has taken place, but prior to the beginning of classes.

2. **Decision-Making Framework, RC-2**

Let us put some numbers to the ingredients. Suppose that the English Department of 10 faculty, to continue with our example, has an average (annual) salary of $70,000 and that each faculty member is required to teach 7 courses (or sections) during the academic year. Then the average cost per course is $10,000. If each student is charged $1,000 tuition for a course, then 10 students are needed to meet the average cost for the course and 15 are needed to satisfy criterion RC-2.

(a) The salary of each department member is needed so that the average salary for that department can be determined.

(b) As to merits, RC-2 is an improvement over RC-1 in that the run/cancel decision for a course is not dependent on the instructor assigned to the course but on the number of students enrolled in it, which makes more sense. RC-2 is clearly easier to implement than RC-1.

As to disadvantages, if the department consisted mostly of relatively high paid senior faculty with an average salary of $105,000, for example, an enrollment of 20 in a course would be needed to satisfy RC-2. If the department consisted mostly of recently hired junior faculty with an average salary of $35,000, let us assume, an enrollment of 10 in a course would be needed to satisfy RC-2. The implementation of RC-2 might vary considerably from department to department depending on the mix of the department's junior and senior faculty.

This does not make good academic sense. Departments are not unconnected units to be considered in isolation, but are intended to service a larger whole - the university itself. The course run/cancel criteria should reflect the needs of the university as a whole and the shortcoming of RC-2 is that is it not designed to do this.

6

Good Data, Good Data My Kingdom for Good Data (Part 3): Can You Trust Polls?

6.1 Preface

The books I examined refer to poll results as sources of data for some of the applications presented. Little or nothing is said about the reliability of this most significant source of data. Political analyst Ben Wattenberg's remark 'One can defend almost any position on almost any subject by referring to public opinion polls'[1] should make clear that the practice of polling deserves careful scrutiny.

---

[1] M. Wheeler, *Lies, Damn Lies, and Statistics: The Manipulation of Public Opinion in America* (New York: W. W. Norton Co. and Dell Pub Co., Inc, 1976, 1977), p. 176.

6.2 How Trustworthy are These Poll Results?

Keep the following in mind:

1. Context and Basic Information

Focusing on poll results without an appropriate context is misleading. News accounts of poll results should give information about the date the poll was taken, sample size, survey design, percentage of respondents among those contacted, response options, random sampling error and what it means. Out-of-context poll results are best viewed with questioning skepticism.

2. Questions.

The complete wording of questions should be provided. We may then ask ourselves: Are the questions clearly posed? In addition to the questions posed, are there questions that should have been posed to avoid overall bias in the questioning? Are there leading questions whose coloring favor a certain kind of response? Are there personal questions that a respondent might be reluctant to answer truthfully? At first thought it might not seem like a big deal to ask a person his preference in an upcoming election, but where an atmosphere of repression or intimidation exists it is indeed a most sensitive question.

3. The Response Rate.

What was the response rate? A major lesson of *The Literary Digest* 1936 presidential poll is that a low response rate may render the results obtained untrustworthy as a basis for predicting the attitudes of the target population. This issue is more complex than it might seem to be at first sight. For further discussion, see G. Langer, "About Response Rates: Some Unresolved Questions," *Public Perspective,* May/June 2003; 16-18.

4. Self-Selected Respondents.

A popular, but seriously flawed, polling technique is the mail-survey, magazine, online, or telephone poll in which the public is invited to respond to a written questionnaire, or call one number to register approval of a candidate or position and another number for disapproval. Polls of this sort lend themselves to gross manipulation by individuals and pressure groups and give us no handle on the opinions of non-respondents. Self-selected respondents represent only themselves.

Any claim or suggestion that results obtained from polls of this sort express the views of a wider population is totally without merit.

5. Online Polls.

The first stage of online polling does not differ from mail-in or telephone-response polls. Pseudo-poll would be a more appropriate label for such techniques rather than poll, which has come to suggest a rigorous, scientific framework not possessed by the first stage of online polls and their ilk.

The advantage of online pseudo-polls is that the internet dimension permits the development of this pseudo-poll into a "legitimate" poll, sometimes termed an interactive online poll. This development is in progress.

When presented with the results of an online poll there is no way of knowing whether they are from a pseudo-poll or interactive online poll unless the methodology is identified. This is often not done.

Beware.

| 6. | Who Commissioned The Poll and Who's Doing It? |

Many honest polls/surveys are commissioned by interested parties, but there are fair-minded interested parties and not-so-fair-minded interested parties who are more concerned with manipulating public opinion than obtaining an objective assessment of it. It is useful to know who wants to know and who's doing the survey to help provide some perspective for the results obtained.

6.3 FOOD FOR THOUGHT

1. **Southchester's School System**

The City Council of Southchester wants to obtain a sense of the public's view of the city's school system. A random sample of 1000 of Southchester's residents was sent a questionnaire. Of the 100 responses received, 32 gave the school system a favorable rating and 68 gave it an unfavorable rating.

Dismayed by this overwhelming show of lack of confidence in Southchester's school system, the City Council voted to dismiss the head of the board of education. Does the reliability of the 68% unfavorable rating support this action being taken?.

2. **Mayor Keith Joos**

Mayor Keith Joos of Masters-on-the-Mississippi is planning to run for re-election. On a recent talk show he invited the public to call a 900 number to express a favorable or unfavorable rating of his administration. Five hundred calls were received; 350 callers gave his honor an unfavorable rating and 150 gave him a favorable rating. With a 70% unfavorable rating, he strongly considered not running for re-election. Is his pessimism over the results of this call-in warranted?

3. **Take Military Action Against Iraq?**

Based on the results of a *Christian Science Monitor*/TIPP poll conducted Oct. 7-13, 2002, it was concluded that "Seventy-five percent of Americans say it's important that the U.S. take military action against Iraq by April." (B. Knickerbocker, "Americans Back Iraq War—Warily," *The Christian Science Monitor*, Oct. 17, 2002; 1.) The question posed and the results were the following:

How important do you think it is for the U.S to take military action within the next 6 months in order to remove Saddam Hussein from power in Iraq?

Very important:	46%
Somewhat important:	29%
Not very important:	13%
Not at all important:	9%
Not sure / Refused:	3%

(a) Do you Agree or Disagree with the view that this being the only question posed concerning Iraq and the nature of its wording rig the poll to favor going to war with Iraq?

(b) If you answered Disagree, explain the basis for your view.

(c) If you answered Agree,

   (i) explain the basis for your view;

   (ii) are there questions and response options that you would recommend be added to the survey to obtain a more accurate assessment of the public's views? If so, state them.

(iii) would you recommend that the wording of the aforenoted question be modified to make it more neutral and less of a leading question? If so, do you have suggestions on how this might be done?

4. **The Center for Critical Thinking**

The Center for Critical Thinking on Domestic and World Affairs is preparing a questionnaire to obtain a sample of public opinion on domestic and world affairs. The following is a sample of some of the questions being posed.

(a) Would you vote for a presidential candidate who was willing to take more out of your pocket by raising taxes?

(b) Do you want the nation's defense capability reduced by budget cuts in an age of rampant terrorism?

(c) Do you believe that very high priority should be given to reducing the crushing budget deficit that has been imposed on our country?

(d) Do you believe that we should continue to squander money on foreign aid while there are so many urgent domestic needs that require attention?

The Center is engaged in pre-testing its questionnaire. To assist them, please address the following questions.

(i) Do you Agree or Disagree with the view that these are biased questions intended to lead the respondent to a favored response?

(ii) If you answered Disagree, explain the basis for your view.

(iii) If you answered Agree,

(a) explain the basis for your view;

(b) how do you recommend that the questions be reworded to eliminate bias and help the Center for Critical Thinking obtain a more accurate assessment of the public's views.

5. **Wording**

How does the choice of wording affect how Americans view Iraqi involvement in Sept. 11? T. Zeller, "How Americans Link Iraq and Sept. 11," *The New York Times,* March 2, 2003; Wk-3.

6.4 ANSWERS/DISCUSSION OF FOOD FOR THOUGHT

1. **Southchester's School System**

The low response rate of 10% to the poll renders unreliable the conclusion that 68% of Southchester's residents view the city's school system in an unfavorable light.

2. **Mayor Keith Joos**

No. He only heard from people who cared enough about the issue or poll to call. The rest of the voters could possibly turn the tide.

3. **Take Military Action Against Iraq?**

   (a) Agree

   (b) Only one option is offered—take military action within six months.

      (c-i) There are, of course, many other possible options that might have been made available.

      (c-ii) For example, how important do you think it is for the U.S. to work with the U.N. in resolving its differences with Iraq; should military action against Iraq be taken by the U.S.? An important response option, not included here, is Need More Information.

(c-iii) I don't see how you can suitably modify the wording of the question posed to make it less of a leading question. To make it more "neutral" it would have to be dropped and replaced by more neutral alternatives or kept with other options included.

The omission of alternatives to talking military action make the question strongly biased. Question: What would the inclusion of a non - military action option have had on the public polled and decision made by those who closely watch polls and are influenced by them?

4. **The Center for Critical Thinking**

(i) Agree.

iii(a) "Take more money out of your pocket" in (a) makes the question biased; it is clear from this phrase that the Center for Critical Thinking does not favor raising taxes and is leading potential respondents to the answer in the same direction.

The phrase "defense capability reduced by budget cuts" makes (b) biased.

The word "crushing" makes (c) biased.

The word "squander" makes (d) biased.

iii(b) Bias can be subdued by dropping the inflammatory phrases or words. For (a) we have: would you vote for a presidential candidate who was willing to raise taxes?

5. **Wording**

'The substitution of just one word, or the order in which questions are asked, can have a profound effect on the results,' "said Mark A. Schulman, the president of the American Association for Public Opinion Research. Asking someone, for instance, whether they think Mr. Hussein 'helped the terrorists' in the Sept. 11 attacks, as the Pew Research did last October [2002], will often yield more positive responses than will questions that ask if Mr. Hussein was 'personally involved' in the attacks, as a Time/CNN poll did at roughly the same time."

7

These Numbers / Statistics are Not Reliable or Not Relevant. So What?

7.1 PREFACE

Consequences of bad numbers/statistics are noted in the preceding chapters, which give us some sense of so what?

A start, but more needs to be said, I submit. If our students and colleagues are to knowledgeable about the seriousness of the matter so that it is given primary attention rather than peripheral notice, attention should be given to cases for which this issue takes primary focus.

7.2 CASES

Case 1: The Consumer Price Index (CPI)

The Consumer Price Index (CPI) value is intended to measure the behavior of inflation. It reflects the average change in prices over time of a market basket of about 80,000 goods and services each month in seven major groups. The CPI takes into account changes in the price of such items as food, clothing, housing, energy, transportation, medical and dental services, medical drugs, and other goods and services that people require for day-to-day living. Different urban areas, housing units and retail establishments around the country are taken into account in compiling this index. Then price changes for the various categories in each region are weighted to take into account the relative amount spent by that particular locale. Finally, local data is combined to obtain an overall average.

The CPI value generated as a valid consequence of the assumptions/postulates of the underlying CPI model is published in the third week of each month by the Bureau of Labor Statistics (BLS). The monthly increments are added to yield an annual CPI figure.

So What?

So what? is answered by looking at the wide reach of the CPI

Fully Indexed Programs. These are programs for which automatic increases in benefits are triggered by increases in the CPI. Social Security, received by about 50 million beneficiaries, is the best known of these programs. Others include railroad retirement, with about 800,000 beneficiaries, supplemental social security income, with about 6.5 million recipients; veterans' compensation; and federal military and civilian employee pensions, paid to about 4 million retirees. The official poverty line rises each year in accord with the behavior of the CPI, which affects about 26 million recipients of food stamps, 25 million in subsidized child nutrition programs, and 5 million with federal student grants.

Taxes. To protect taxpayers from the effects of inflation taxes are adjusted in a number of ways. This includes tax brackets, which determine tax rates on income; personal exemption and standard deduction levels; earned income tax credit; limit on itemized deductions; and pension contribution limits.

Economic Statistics. Real Growth in the Gross Domestic Product (GDP) and productivity (i.e. growth adjusted for inflation) depends on the CPI.

The smaller the increase in the CPI, the smaller will be the additional benefits paid to beneficiaries, the smaller will be the cost to the government, the greater will be the tax obligation of taxpayers and government tax revenue, and the larger will be the GDP and productivity figures. While there is no such thing as a "perfect" measure of inflation, a good deal rides on making the CPI model as realistic a model of inflation as possible.

See Ch. 12 for discussion of the CPI model which underlies the CPI value.

Case 2. Is Balance the Budget and Pay Off the National Debt the Road to Prosperity?

The issue was reignited when President Clinton proclaimed, "Let's make America debt-free for the first time since 1835," in his January 2000 State of the Union Message. Eliminating the debt shouldn't be the nation's goal, critics responded. The ratio of public debt to Gross Domestic Product, which has been falling for several years, is the number to watch, they countered. Vice President Gore, seeking his party's nomination for President, jumped on the bandwagon by declaring that we should continue to pay down the debt even when the economy slows.

Professor Frederick C. Thayer notes that there were six major periods in American history that began with budget balancing and debt reduction and ended with depression [Thayer; 9].

- 1817-21: The national debt was reduced by 29%. The first acknowledged major depression began in 1819.

- 1823-36: The national debt was reduced by 99.7% to $38,000, a virtual wipeout as Thayer put it. A major depression began in 1837.

- 1852-57: The national debt was reduced by 59%. A major depression began in 1857.

- 1867-73: The national debt was reduced by 27%. A major depression began in 1873.

- 1880-93: The national debt was reduced by 57%. A major depression began in 1893.

- 1920-30: The national debt was reduced by 36%. The sixth major depression, the Great Depression, as it came to be called, began in 1929.

These data by themselves do not "prove" that balancing the budget and reducing debt cause depressions, but they do make clear that these measures are not sufficient to ward off depressions and suggest that if carried out as a first priority, no matter what the circumstances, may serve to turn a recession into a depression. The Great Depression provides some noteworthy lessons. A recession had begun to take hold in the Spring of 1929, followed by the Wall Street crash in October—a bad downturn that was severely aggravated by bad monetary and fiscal policies. In 1932, at the worst possible time, President Hoover raised taxes to help balance the budget and "restore confidence." This was not a mindless decision, but rather part of the orthodoxy of the time. It should give us pause to reflect on the damage implementation of flawed orthodoxy can do.

So What?

Big money is at stake and how it is spent will affect the lives of millions of people. If the first priority is to pay off the debt, other needs—education, health, infrastructure, tax relief, you name it—must take lower priority.

Case 3: The Drug Testing Balloon That Popped

Dr. Robert Fiddes was on the top of the world. He had gone from successful medical practitioner to founder and director of the Southern California Research Institute, a company that tested the effectiveness and safety of new drugs for pharmaceutical companies and brought in millions of dollars. Slightly behind this stunning facade of success and prosperity, research fraud of almost unimaginable proportions was in progress.

It wasn't simply a matter of altering numerical readings, although this was done too, but of compromising the entire framework for obtaining meaningful readings. Blood pressure readings were indeed fabricated, but fictitious patients were invented, medical records were falsified, medical tests were compromised, patients who did not satisfy the criteria for inclusion in a study were enrolled anyway.

The balloon was punctured on February 16, 1997 when federal agents occupied the Southern California Research Institute's Office and confiscated box loads of incriminating documents, leaving in its wake compromised drug study results for almost every pharmaceutical house in the business [Eichenwald and Kolata; 2].

So What?

The basic underlying issue centers on the integrity of the data - which translates to integrity of the numbers generated from the studies. At the end of the road we all pay part of the tab which may come in the form of monetary cost (in many ways the simplest of the costs), quality of life for ourselves, family, friends, members of our community, and life itself.

Case 4. A Faulty, No Fault Divorce Statistic

In her 1985 book, *The Divorce Revolution,* Stanford University sociologist Lenore Weitzman stated that an economic consequence of no-fault divorce is that women's standard of living drops an average of 73 percent in the first year after divorce, whereas men's rises 42 percent. Although sociologists challenged these figures almost immediately, they were the figures that stuck. They were cited in more than 100 national magazines and newspapers, including *The New York Times* and *Cosmo,* at Congressional hearings, in more than 250 law review articles, in at least 24 state appellate and Supreme Court cases, and by the Supreme Court itself. [Faludi; 3] They were the figures of choice of critics of no-fault divorces seeking to overturn or block no-fault divorce legislation. The fact that they were inconsistent with other findings and had not been verified mattered little, if at all. A hot statistic was too good not to be true.

Weitzman's figures proved to be false. In 1993 sociologist Richard Peterson obtained access to Weitzman's data, reanalyzed them using the same methods she had employed, and obtained figures of 27 percent for the decline in the standard of living of women and 10 percent for the rise in standard of living of men. These results were in-line with those that had been obtained by other researchers. Weitzman believes that her erroneous results were due to problems with her computer files.

So What?

Peterson's observations that "the discussion of no-fault divorce and other legal reforms has been seriously distorted by . . . inaccurately large estimates. To be effective, these reforms must be based on reliable data," should be taken seriously for their general applicability.

Case 5. Proper Funding for the Data Gatherers

The problem is that obtaining sufficient funding to do a proper job of number crunching has become an annual battle with Congress.

So What?

Murray Weidenbaum, chief economic advisor to President Reagan, was prompted to observe [Weidenbaum; 10].

The real costs resulting from providing inadequate funding for one of the most useful parts of the bureaucracy—the data gatherers—are truly awesome. An inaccurately high report of inflation can trigger an avoidable policy of monetary restraint followed by needless declines in capital formation, production, and employment. Bad information on productivity can generate investor decisions out of sync with real trends in the marketplace.

Underreporting exports can produce news of "record" trade deficits followed by adoption of tougher protectionist policies, which could undermine the growth prospects for the US as well as those nations it trades with.

Case 6. Hayes vs. Tilden, 1876

The 1876 contest was between two respected men, Samuel Tilden, the Democratic governor of New York, who had broken up the Tammany Hall corruption system, and Rutherford Hayes, the Republican governor of Ohio. Despite their personal standing, the election was exceptionally dirty. Tilden was called a syphilitic swindler and Hayes was accused of murdering his mother in a fit of insanity—an impressive double calumny.

Tilden won 51% of the popular vote and came just one short of a majority in the electoral college. But the votes in three southern states were disputed (Florida was one), and all three eventually sent competing returns to Congress. The House and Senate, however, were controlled by different parties and could not agree on which votes to certify. So they set up a bipartisan commission—seven Democrats, seven Republicans, one independent—to settle their disputes. The independent was then elected to the Senate, which made him ineligible, and his place was taken by a Republican whom Democrats thought would be non-partisan.

When the electoral roll-call began, however, he voted with his party on every occasion. Each southern elector was challenged and, each time, the commission to which the dispute was referred split eight to seven for the Republicans. Hayes was elected amid universal accusations of fraud and sporadic violence. . . .

To placate aggrieved southern Democrats, Hayes agreed to remove the so-called "carpetbagging" Republican governments that had been imposed on Southern States after the Civil War. For their part, Southern Democrats promised to protect the interests of blacks in their states; but they soon broke that promise.

So What?

The real consequence of the disputed 1876 election was the end of the post-war period of civil rights and the start of renewed oppression of southern blacks."[*The Economist*; 1]

Case 7. Soviet Defense Outlays?

The disintegration of the Soviet Union has left the United States facing some critical choices. Answers to questions about the extent to which the United States should scale back military spending are strongly influenced by analyses of the Central Intelligence Agency which indicate that the former Soviet Union had been substantially outspending the United States on military programs for the last 15 to 20 years. America's massive military buildup, especially that carried out during the Reagan years, was predicated on these assessments.

A leading American analyst of Soviet military spending, Franklyn D. Holzman of Tufts University, has consistently taken issue with the assumptions and procedures which underlie the C.I.A.'s analysis and conclusions. Writing in 1979, Holzman notes [Holzman; 5]:

There are many sources of possible exaggeration in the C.I.A. estimates of Soviet military expenditure relative to America's. Three of them follow:

1. Comparisons of military outlays can be made either in dollars or rubles. The C.I.A.'s published comparisons are always in dollars. Prices expressed in dollars exaggerate Soviet expenditures. This is because the Soviet armed forces have twice the personnel of America's but add only a little more new equipment each year, and

because, in the words of the Director of Central Intelligence, Adm. Stansfield Turner: "In the United States manpower is relatively more expensive than hardware [while] in the Soviet Union military hardware is much more expensive than manpower." So, when the cost of the personnel of the Soviet armed forces, with their 4.5 million people—the precise number is hard to ascertain—is valued at American armed forces wages, a high Soviet defense figure, in dollar terms, results. This figure would be about $10 billion smaller if military pay were adjusted for the lower educational and training levels of Soviet soldiers. A 20 percent pay discount is regularly made by the C.I.A. in dollar comparisons involving other sectors.

2. A ruble comparison, which the C.I.A says is as valid as the dollar comparison, exaggerates American expenditures. This is because our armed forces have more equipment per person than the Soviet forces and because equipment is relatively high priced in the Soviet Union. The C.I.A. admits this and in response to congressional questioning presented an unofficial comparison in rubles that put Soviet 1977 defense expenditures at 25 percent more than America's. This is less of a difference than the official dollar comparison, which has the Russians outspending us by 40 percent. While these two not-very-different figures satisfy Congressional interrogators, it did not satisfy economists used to such United States Soviet comparisons. Experience has shown that ruble-dollar differentials typically exceed 50 percent. Clearly, then, if the Soviet Union outspends the United States in dollars by 40 percent, one would expect the United States to equal or outspend the Russians in rubles. These C.I.A. figures, therefore, are highly suspect.

3. According to the C.I.A., the major reason why a careful ruble estimate is not made and published is that while all military equipment the Russians produce is within our technology and can be given a real dollar price, a large part of United States equipment is beyond Soviet technology and cannot be given an actual ruble price. The C.I.A. procedure in valuing American high-technology equipment is to use ruble prices "applicable to the closest substitute goods which can be produced in both economies." What this means

is that the C.I.A.'s ruble calculation values this American equipment at ordinarily high Soviet ruble prices but not at what the former Director of Central Intelligence William E. Colby called prices that are so high as to be "almost uncountable." No wonder American defense expenditures priced in rubles are estimated at less than the Russians' defense expenditure. If a properly high ruble price tag could be put on our high technology, the American defense package would certainly cost the Russians more to produce than their own. It might well be that they cannot produce our defense package at any cost.

The major fallacy in the C.I.A. procedure is that the very dimension of the arms race in which America has the greatest advantage-advanced technology—and which makes most of the difference between military superiority and inferiority, is enormously undervalued.

In an analysis published in 1989 Holzman observes[Holzman; 6]:

> ... over the last 15 years, the C.I.A.'s estimates have been riddled with errors and misrepresentations, all making Soviet defense expenditures appear larger than they actually are.
>
> From 1975 through 1983, the C.I.A. said Soviet military spending was increasing 4 percent to 5 percent a year. In 1984, the agency acknowledged that it had been wrong—that the increase since 1975 had been only 2 percent a year. After conceding its error, however, the C.I.A. failed to make the appropriate adjustments in its Soviet military spending figures.
>
> In 1982, the Soviets instituted a wide-ranging price reform, the first in 15 years. In 1986, the C.I.A. switched its estimates of Soviet defense spending, which had been in 1976 ruble prices, into the new 1982 ruble prices. Soviet defense spending took a large leap upward because, according to the C.I.A., prices of weapons had increased much faster than prices of civilian goods between 1970 and 1982. But this was all guesswork.

In November, 1987, the agency conceded that it had been unable to obtain 1982 weapons prices and, further, that there was reason to believe that weapons prices had risen more slowly than prices of civilian goods. Yet the new and much higher figures for Soviet defense expenditures were never rescinded.

In 1986, the C.I.A. made another change in established practice. Without explanation or warning, it began to include in its single estimate of Soviet military spending a number of civilian activities that are not included in estimates of United States military spending (such as civilian defense, internal security forces and civilian space activities). This made Soviet defense appear still larger relative to United States defense than it should have been.

Holzman concluded that when the C.I.A.'s errors and assumptions are corrected Soviet military spending as a share of gross national product falls almost 50 percent, from 16 percent claimed by the C.I.A. to 9 percent.

During the confirmation hearings held in the Fall of 1991 on the fitness of Robert Gates to be C.I.A. Director, senior analyst Marvin Goodman and two other C.I.A. analysts testified that while serving as Deputy Director of Intelligence Gates tailored intelligence estimates to suit Administration policy in several important areas. One of the areas cited was concerned with Soviet capabilities and intentions.

> In connection with this state of affairs Holzman comments [Holzman; 7]:
>
> Professor Goodman's revelations come as no surprise to me; indeed, they suggest an answer to puzzles raised over the last 15 years in my studies of C.I.A. estimates of Soviet military spending.
>
> The C.I.A. estimates have been, in my opinion, continually slanted upward to make Soviet military spending appear larger than careful analysis of the data suggests, thereby supporting the Reagan-Bush-Pentagon military buildup policies.

It was clear to me that the decisions to slant the conclusions were made at the highest levels because the techniques used

were so unacceptable on scholarly grounds. In my work on military spending, I have come to know many of the C.I.A. analysts who prepare these estimates and I believe that, had they been free of political pressure, the estimates would not have been distorted.

So What?

As to the implications of these developments for the post cold war military budget, Holzman argues:[Holzman; 8]:

> Had there not been a C.I.A. error followed by a cover-up, our military spending might have continued to increase at a slower rate. It is estimated that for the decade 1979-1988 the United States would have spent on defense approximately $800 billion less in present-day prices. . . . The Pentagon used exaggerated estimates of Soviet military spending to help get enormous budget increases. Now that the cold war is over, this overspending should be taken into account in evaluating the Pentagon's claims to severely stretched resources.

References

1. *The Economist,* "Watch yourself at Dinner, Dubya," Nov. 29, 2000 These details are culled from Norman Omstein's history of disputed elections contained in "After the People Vote", a recently reissued definitive guide to the electoral college and its constitutional provisions (AEI Press).

2. K. Eichenwald and G. Kolata, "A Doctor's Drug Studies Turn into Fraud," *The New York Times,* May 17, 1999.

3. S. Faludi, "Statistically Challenged," *The Nation,* April 15, 1996.

4. M. Gardner, "The Power of Statistics to Affect Lives—Even When They're Wrong," *The Christian Science Monitor,* May 2, 1996.

5. F. Holzman, "Of Dollars and Rubles," *The New York Times,* Oct. 26, 1979.

6. F. Holzman, "How C.I.A. Concocts Soviet Defense Numbers," *The New York Times,* Oct 25, 1989.

7. F. Holzman, "How C.I.A. Invented Soviet Military Monster," *The New York Times,* Oct. 3, 1991.

8. F. Holzman, "C.I.A. Error Still Bloats Our Military Budget," *The New York Times,* Feb. 2, 1993.

9. F. C. Thayer, "Do Balanced Budgets Cause Depressions," *The Washington Spectater,* Jan 1, 1996.

10. M. Weidenbaum, "Fund the Number Crunchers," *The Christian Science Monitor*, Sept. 16, 1999.

8

Are Quantitative Studies Preferable to Qualitative Ones? Can They Be Combined?

8.1 Preface

The sample of books I examined gives attention to qualitative studies, but the questions raised in this chapter's title receive no attention. Since finding a discussion might prove difficult, I thought it would be useful to present the discussion that my colleagues and I present in our book, with some refinements.

8.2 Sexuality By the Numbers, or Not?

In 1987 Shere Hite published *Women and Love: A Cultural Revolution,* her third book on human sexuality. In her first two books, *The Hite Report* (1976) and *The Hite Report on Male Sexuality* (1981), Hite restricted herself to telling what she had learned from women and men who replied to the extensive questionnaires concerning their sexual problems and attitudes she had circulated. For her first book she circulated approximately 100,000 questionnaires, from which she received 3019 responses for a response rate of about 3%. Four different versions of the questionnaire were sent to women's organizations that were asked to circulate them. A similar methodology was employed for her second book. Approximately 119,000 questionnaires were distributed with a response rate just under 6% being obtained.

Hite was not claiming that her sample was representative of women and men in general. Hers was a qualitative rather than statistical study. In statistical studies, the same questions must be asked of all prospective respondents with the same response options being available to them all. Uniformity of the underlying conditions and the choosing of a "representative" sample so that the results obtained could be projected onto the population at large are essential for a statistical study. Qualitative studies, on the other had, focus on the special qualities of each individual potential respondent. Capturing the diversity inherent in individuals takes priority over ensuring uniform underlying conditions. It is not a matter of one kind of study being superior to the other, but rather of which methodology is appropriate to the study being undertaken. Although her first two books raised much controversy, Hite was on safe methodological ground.

In her third book, Hite attempted to cross the bridge from the qualitative results she had obtained to statistical generalizations about sexual attitudes of women in America. The bridge collapsed. Her methodology was almost universally criticized. ABC News in conjunction with *The Washington Post* conducted a telephone poll for October 15-19, 1987 to see if they could duplicate her results. They could not; their results were sharply at variance with those projected by Hite.[1] To take two examples, she found 84% of women as not being satisfied emotionally with their relationships; ABC/WP found 7% of married women and single women in a relationship as not being emotionally satisfied. Hite found 78% of women feeling they are only occasionally treated as equals most of the time. The ABC/WP figure was 9%. There were differences in the way questions were posed, but they were not startling. As to who is closer to the mark, ABC/WP clearly takes the Trustworthy Prize because of its sound statistical methodology.

---

[1] "Hite/ABC Poll Comparison Analysis," news release by ABC for 6:30 pm, EST, Mon., Oct 26, 1987.

9

You Want More Accuracy in Your Figures? Add More Decimals to Them?

9.1 PREFACE

This is not an issue that is mentioned in the basic statistics texts that I examined. My experience is that my students, almost without exception, believe that to obtain greater accuracy add to the flow of decimals. While this is correct for numbers such as $\sqrt{2}$ and π it is not correct for numbers that arise in applications and are rounded off, a distinction they are not prepared to make unless the issue is brought to their attention.

M. Richardson in his book[1] gives the following illustration:

Take a tub of water and place 7 empty pails on the floor. Spill the water from the tub into the pails, giving each an equal share, as nearly as possible. It would be appropriate to say that each pail contains roughly one-seventh of the tub of water. While this may not seem very accurate, it's the best that can be said. If, however, we divide 1 by 7 to obtain 0.142857 . . . , and proclaim that each pail contains about 14.2857 percent of the water in the tub, one might be deluded into believing that this is a more accurate description of the amount of water in each pail if one equates accuracy with number of decimal places. From this point of view the accuracy of the description would be further enhanced by carrying the division of 1 by 7 to still more places, 0.1428571428571 . . . , for example. It's nonsense, of course; we are no better off than when we said that each pail contains about 1/7 of the tub of water. The important point is, real accuracy depends on the accuracy of the data.

---

[1] M. Richardson, *Fundamentals of Mathematics*, Revised Edition. (New York: The Macmillan Co., 1958), 402.

For a comprehensive discussion I recommend Freund and Williams[2],

9.2 FOOD FOR THOUGHT

1. **Find the Sum of 2.34 and 1.131,**

    (a) assuming they are exact,
    (b) assuming they are approximate numbers with 2.34 known to two places and 1.131 known to three places.

2. **Find $\sqrt{2}$,**

    (a) assuming 2 is exact,
    (b) assuming 2 is considered approximate,

3. **The "Heavy" *Basic Statistics* Text.**

A recent edition of *Basic Statistics* was the focus of a wrist strength test at Ecap University. Table 9.1 gives the time lengths, in minutes, that the

---

[2] J. Freund, F. Williams *Modern Business Statistics* (rev. by B. Perles, C. Sullivan, (Prentice-Hall, 1969), Appendix III, Calculations with Rounded Numbers, 492 - 4.

population of 156 students at Ecap University who used the text were able to hold it in reading position in one sitting before their wrists gave out.

Table 9.1

1.7	4.3	6.0	4.7	1.7	7.2	3.1	3.9	3.3	7.0	2.5	4.9	5.2
2.3	4.5	6.9	3.4	3.8	6.6	3.6	3.5	5.5	5.4	5.9	6.1	4.4
7.9	4.7	5.1	6.0	5.9	8.4	5.1	6.3	5.2	4.4	5.2	3.0	4.4
4.3	3.3	5.0	3.8	7.0	2.9	3.0	4.1	4.5	5.2	3.3	4.2	2.5
5.1	2.1	4.6	2.9	4.1	5.9	4.7	4.8	4.1	5.5	5.0	2.6	4.4
6.1	5.0	6.3	4.6	8.1	4.6	6.1	6.6	3.2	5.9	1.8	3.2	1.5
5.7	7.3	4.3	3.2	3.4	4.1	5.0	1.7	4.3	4.2	3.5	4.1	4.5
4.8	3.7	3.0	4.9	7.2	4.5	4.9	6.5	3.1	5.2	4.8	3.9	4.9
7.5	3.8	4.5	6.8	4.1	0.3	5.9	2.4	3.6	3.4	1.9	3.4	1.6
3.5	3.4	3.6	2.6	5.9	5.1	4.9	6.2	5.0	3.9	6.1	5.6	4.6
7.5	6.7	4.4	5.8	4.5	5.1	5.3	4.4	5.8	4.7	2.5	3.5	7.6
4.3	4.7	5.1	5.8	3.1	2.3	5.7	8.2	0.8	2.5	7.2	2.6	6.7

Determine the mean μ.

9.3 ANSWERS/DISCUSSION OF FOOD FOR THOUGHT

1. **Find the Sum of 2.34 and 1.131,**

    (a) assuming they are exact:
    2.34 + 1.131 = 3.471; all places are retained

    (b) assuming they are approximate numbers with 2.34 known to two places and 1.131 known to three places.

Their sum should be recorded as 3.47 (to two places) rather than 3.471. The number 3.471 would give a misleading sense of accuracy.

2. **Find $\sqrt{2}$,**

   (a) assuming 2 is exact:

If 2 is considered to be exact, then the accuracy of an approximation of $\sqrt{2}$ is increased by expressing its value in terms of more decimal places.

$$\sqrt{2} \approx 1.41$$

More accurate: $\sqrt{2} \approx 1.414$
Still more accurate: $\sqrt{2} \approx 1.414214$
How accurate an approximation of $\sqrt{2}$ should be taken in a computation situation depends on the setting of the problem.

   (b) assuming 2 is considered approximate;

If 2 is considered approximate, the accuracy of an approximation of its square root is not enhanced by taking more and more decimal places.

3. **The "Heavy" *Basic Statistics* Text**

Determine the mean μ.

The holding-time numbers arising from the "Heavy" *Basic Statistics* text situation are approximate values correct to one place. The result of a series of computations with such data cannot be expected to be accurate beyond one place. In carrying-out the computations it is advisable to carry one more place and round off at the end.

My colleague Irwin Kabus notes that, "this counsel, as with all advice, should not be followed blindly, but take circumstances into consideration and be exercised with discretion. If your pet cat, for example, weighed in at 5 pounds and then at 6 pounds the following month (approximate values), it makes good sense to record its average weight as 5.5 pounds. To round off this average to 6 pounds would give us a highly distorted view of the underlying situation. On the other hand, if your pet pony weighed in at 500 pounds and later at 501 pounds, for an average weight of 500.5 pounds, considering the magnitude of these values, it would not distort the sense of the situation to round off the average to 501 pounds."

Concerning the "Heavy" *Basic Statistics* text:

$\Sigma x = 714.0$ and $N = 156$.

$$\mu = \frac{714.0}{156} = 4.6$$

10

Statistics Tea Leaves: What Do These Statistics Tell Us? Their Limitations?

10.1 Preface

How statistics are interpreted will, at best, depend on the judgment and capacity of the interpreters and, at worst, on the cleverness of the spin doctors entrusted with putting on them the best possible spin. It is often the case that "experts" see very different things in the same statistics and that different statistics concerning the same issue seem to be contradictory.

This dimension is not mentioned in the sample of basic statistics books I examined.

10.2 CASES

Case 1. There's Less to Baseball Statistics Than Meets the Bat

In his book *The Last Yankee: The Turbulent Life of Billy Martin,* David Falkner [Falkner; 2] concludes that Martin was the best manager of his era, possibly of many eras. Falkner's judgment was strongly influenced by baseball statistics compiled by the Elias Sports Bureau and a formula which claims to show which managers' teams won more games than they were reasonably expected to win.

In his review of Falkner's book George F. Will [Will; 6; p. 17] disputes Falkner's conclusion which, he argues, the rest of the book refutes. "In fact," Will notes:

> The Last Yankee might usefully be made required reading for graduate students in the social sciences and all others who need to be immunized against the seduction of numbers . . . There are limits - and Mr. Falkner's reporting shows that Elias passed them regarding Martin - to the ability to capture messy reality in tidy formulas.

Case 2. How Successful Was the Patriot?

During the Gulf War television viewers were moved by scenes of Patriot missiles streaking across the sky to intercept Iraqui Scud missiles that had been launched against Israel and Saudi Arabia. The Patriot's success seemed to epitomize the success of a high tech, low causality military campaign.

Apart from very successful military public relations, how successful was the patriot as a military tool? Different statistics have been given and, as is almost always the case, it is important to look further than the statistics if a realistic picture is to emerge.

The Army originally stated that Patriots "intercepted" 45 of 47 incoming Scud missiles, and President Bush revised that to 41 of 42. What does this mean? Brigadier General Robert Drolet of the Army's Missile Command testified that "a Patriot and a Scud passed in the sky." There are other statistics of interest. Before Patriots were employed in Israel, 13 Scuds fell near Tel Aviv. There were no deaths, but 115 people were wounded and 2,698 apartments were damaged. After Patriots were employed in this region, 11 Scud attacks left 1 dead, 168 injured and 7,778 apartments damaged. (see [Jagger; 3] and [Marshall; 5]) This is explained by the fact that successful hits led to more deadly debris being sprayed over a larger area than otherwise would have been the case and that the Patriots tended to strike the bodies of Scuds, leaving their warheads armed and able to cause significant damage on landing. But then it should also be kept in mind that the Patriot was not designed as an antimissile weapon, but to defend against fast-flying aircraft.

Case 3. The Reagan Economic Boom: Blessing or Disaster?

Martin Anderson, former advisor to President Reagan and senior fellow at the Hoover Institution, employs statistics to support his view that the Reagan economic boom was the greatest ever [Anderson; 1]:

Anderson's View

The two key measures that mark a depression or expansion are jobs and production. Let's look at the records that were set.

Creation of Jobs

From November 1982, when President Ronald Reagan's new economic program was beginning to take effect, to November 1989, 18.7 million new jobs were created. It was a world record: . . . The new jobs covered the entire spectrum of work, and more than half of them paid more than $20,000 a year. As total employment grew to 119.5 million, the rate of unemployment fell to slightly over 5 percent, the lowest level in 15 years.

Creation of Wealth

The amount of wealth produced during this seven year period was stupendous - some $30 trillion worth of goods and services. Again, it was a world record ... According to a recent study, net asset values - including stocks, bonds and real estate went up by more than $5 trillion between 1982 and 1989, an increase of roughly 50 percent ...

Income Tax Rates, Interest Rates and Inflation

Under President Reagan, top personal income tax rates were lowered dramatically from 70 percent to 28 percent. This policy change was the prime force behind the record breaking economic expansion ...

The Stock Market

Perhaps the key indicator of an economy's booms and busts is the stock market, the bottom line economic report card ... starting in late 1982, just as Reaganomics began to work, the stock market took off like a giant skyrocket. Since then, the Standard & Poor's index has soared, reaching a record high of 360, almost triple what it was in 1982.

There were other consequences of the expansion. Annual Federal spending on public housing and welfare, and on Social Security, Medicare and health all increased by billions of dollars. The poverty rate has fallen steadily since 1983.

When you add up the record of the Reagan years, and the first year of President Bush ... the conclusion is clear, inescapable and stunning. We have just witnessed America's Great Expansion.

Leontief's View

In a reply, Nobel Price winning economist Wassily Leontief [Leontief; 4] concedes some of Anderson's statistics but goes on to look at a number of cost thorns in his statistical rose garden.

True, the long recovery from the deep depression that brought President Reagan to power carried this country to the high point of the usual cyclical wave characterized by a low rate of unemployment and a high gross national product. It is most likely that wholesale tax cuts inaugurated by Mr. Reagan have made the level of the G.N.P., as measured by the Government statisticians, several billion dollars higher than it would otherwise have been. But at what a cost.!

Drastic cuts in public spending (except for military purposes) left the physical infrastructure of this country in ruin. City streets and transportation facilities, water-supply and sewage systems, particularly in large metropolitan areas, are collapsing, the once glorious interstate highways are crumbling, and cramped airports are incapable of handling the rapidly increasing traffic. Despite the valiant effort of the underfinanced, underpowered Environmental Protection Agency, our lakes, rivers and forests are succumbing to deadly acid rain.

What is even worse, the intellectual, cultural and social infrastructure of the country has suffered even more during this greater-than-ever boom than its physical counterpart. Primary and secondary schooling have been so weakened that a whole generation of boys and girls can hardly read, write or count, while the soaring price of higher education makes it impossible for many young people to take advantage of it.

No wonder the competitiveness of the United States is rapidly declining; many of our high technology industries are losing one battle after another in the struggle for their share of the foreign and even their own domestic market. At the same time, the rich are getting richer, and the poor are getting homeless.

Let us hope that contrary to Mr. Anderson's expectations the "Reagan boom" will not continue in its present form for four or eight more years. If it does, the United States will find itself entering the 21st century as the richest country (in total value of stocks and bonds traded on the stock exchanges), but culturally and socially less advanced than other developed countries.

References

1. M. Anderson, "The Reagan Boom - Greatest Ever," *The New York Times,* Jan. 17, 1990.

2. D. Falkner, *The Last Yankee: The Turbulent Life of Billy Martin* (New York: Simon & Schuster, 1991).

3. J. Jagger, "Why Patriot Didn't Work as Advertised," *The New York Times,* June 9, 1991.

4. W. Leontief, "We Can't Take More of This 'Reagan Boom,'" *The New York Times,* Feb. 4, 1990.

5. E. Marshall, "Patriot's Scuds Busting Record is Challenged," *Science,* May 3, 1991.

6. G. Will, "Paranoid in Pinstripes," Review of [2], *The New York Times Book Review,* April 5, 1992.

10.3 FOOD FOR THOUGHT

1. **Discrimination or Difference?**

Concerned about charges of subtle patterns of bias at its executive levels, The United Federation of Worlds set up a Commission to investigate. During its hearings Lork from Mork pointed out that while Morkians are 30% of the Federation's work force at its lower levels, they make up only 1% of its executive staff. "Good faith recruitment efforts have been made," observed Lork, "but the stastistics show subtle patterns of discrimination against Morkians." Tallia from Talos I disagreed. "The statistics show discrepancies," countered Tallia. "The subtle patterns of discrimination are your interpretation of the discrepancies.

   (a) Who is right?
   (b) Are there other explanations that might account for the discrepancies?

2. **What is the Significance of a Grade of 50?**

Ellen Ames and Ann O'Neil, professors of economics at Huxley College, saw a grade of 50 on Professor Ames's last economics exam in different terms. "A grade of 50 is an F," noted Professor Ames. "But it's the highest grade in the class," replied Professor O'Neil, "and as such an F does not make sense." Discuss the significance of 50 as the highest grade on an exam, and as an F.

3. **Plutonians are a Bad Lot?**

"80 percent of the crimes in this city were committed by Plutonians," remarked Oscar to his wife Janet. "They're a bad lot." Are there other interpretations of this figure? Are there other figures that might be relevant? Discuss.

4. **Cause for Pessimism or Optimism?**

One thousand senior high school students in Ralph City took a college level math course, with 80% of them passing the statewide standard exam. The following year 3000 senior high school students in Ralph City took the course, with 50% of them passing the statewide standard exam. Do these figures give us cause for pessimism or optimism concerning Ralph City's education system?

5. **Tallia vs. Lork: Part 2**

Continuing their discussion of what statistics show, Tallia brought up the statistics concerning the Universe Games. "Consider the Universe Games held every hundred years. For the last two thousand years 75% of the participants chosen by the trials have been Morkians. Does this mean that the trials were biased in favor of the Morkians?" "Certainly not," answered Lork, "they earned the right to be there in the trials that were held."

   (a) Do the statistics mean that the trials were biased in favor of Morkians?

   (b) Are there other "explanations" that might account for the statistics?

10.4 Answers/Discussion of Food for Thought

1. **Discrimination or Difference?**

    (a) Who is right?

We would have to agree with Tallia that the statistics show discrepancies. How the discrepancies are interpreted is another matter. Lork is offering one interpretation of the discrepancies that the statistics reveal.

    (b) Are there other explanations might account for the discrepancies?

In general Morkians do not have the education to qualify for positions at the executive level; or, Morkian culture gives high priority to sports and arts and sciences and Morkians tend not to be interested in the kind of positions available at the executive level.

Note that Morkians and Talosians may be replaced by two contending groups of your choice on Earth.

2. **What is the Significance of a Grade of 50?**

Prof. Ames reflects the view that the exam was appropriate to what was taught and that the students did not study enough. Prof. O'Neil reflects the view that the exam was too difficult considering what was emphasized in the course and that 50 is a reflection of the difficulty of the exam rather than insufficient student study effort.

3. **Plutonians are a Bad Lot?**

The police force, consisting primarily of non-Plutonians, are prejudiced against Plutonians and the 80% is a reflection of this prejudice.

If the Plutonian population level of the community were 80% or higher, the 80% crime level would not be considered excessive.

Note that Plutonian may be replaced by any group of your choice on Earth.

4. **Cause for Pessimism or Optimism?**

> Do these figures give as cause for pessimism or optimism concerning Ralph City's education system?

Optimism: More students took the math course (3000 vs. 1000).

Pessimism: In percentage terms, the percent passing was lower (50% vs. 80%).

5. **Tallia vs. Lork: Part 2**

> (a) Do the statistics mean that the trials were biased in favor of Morkians?

(b) Are there other "explanations" that might account for the statistics?

(a) The statistics show discrepancies. The Morkians earned the right to he there is one interpretation of the discrepancies.

(b) Non-Morkian cultures give much lower priority to the Universe Games than does Morkian culture.

11

Data Scales: Nominal, Ordinal, Interval, Ratio: Caution Advised.

11.1 PREFACE

Data scales receive shortshrift in the sample of basic statistics texts I examined. Descriptions are given but there is little or no follow-up, a serious shortcoming because the statistical method that may be employed to analyze the data depends on the data scale.

My experience is that students find this topic difficult. The appropriate way to help them overcome their difficulties is to provide illustrations and discussion.

It is my hope that you find the appreach my colleagues and I have employed useful food for thought.

11.2 DATA SCALES

At the end of the academic year each student at Ecap University is asked by the student government to fill out a questionnaire. The following are examples of some of the questions asked.

Personal Data: Please indicate the following:

1. Sex: (1) Male, (2) Female

2. Marital status: (1) Married, (2) Single, (3) Separated, (4) Divorced

3. Age: (1) Under 18, (2) 18-19, (3) 20-24, (4) Over 24

4. Work status: (1) Employed full time, (2) Employed part-time, (3) Not employed

5. Highest SAT score (total): (1) Under 900, (2) 900-1000, (3) 1001-1100, (4) Over 1100.

University Related Data

6. Class designation: (1) Freshman, (2) Sophomore, (3) Junior, (4) Senior

7. Area of major: (1) Business, (2) Humanities, (3) Science, (4) Computer Science, (5) Mathematics, (6) Education, (7) Engineering, (8) Undecided

8. Distance of residence from campus: (1) Under 1 mile, (2) 1-3 miles, (3) Over 3 miles

9. Preferred class room temperature (Fahrenheit): (1) 66-68, (2) 69-70, (3) 71-72 (4) 73-74

10. Satisfaction with teaching quality this past academic year (largest number indicates highest degree of satisfaction): 5 4 3 2 1

The data arising from questions 1, 2, 4, and 7 are said to have a **nominal scale** because the responses express non-overlapping categories for which no ordering is implied. In question 7, for example, eight categories are cited, but there is no ordering implied as stated which would put any one category ahead of any other. Numbers are assigned to the categories to help record the responses, but this is their only role. Nominal scaling is the weakest form of measurement.

1.	The **nominal** measurement level classifies data into non-overlapping categories on which no ranking or order is imposed. Each datum falls into one of the categories.

Suppose we obtain the information from 400 students in response to question 2 shown in Table 11.1.

Table 11.1

Category	No. of Responses
Married	100
Single	240
Separated	40
Divorced	20

If of interest, we could calculate proportions for data on a nominal scale to determine the percentage of respondents in each category. Twenty-five percent of the responding students are married, 60% are single, 10% are separated, and 5% are divorced.

Suppose that the first five respondents answered (1), (1), (2), (4), (2) to question 2. While we could take the average of the numbers 1, 1, 2, 4, and 2, to obtain 2, it would be meaningless to do so in terms of the background situation. To say that the average respondent is single makes no sense.

> We can compute the proportion of the total who are in each category arising from a nominal scale. It's meaningless to compute the average of the numbers used to designate the categories themselves.

The data arising from questions 6 and 10 are said to have an **ordinal scale** because there is an implied order which allows us to speak about one category being better or preferable to another. In question 10 a rating of 5 is better than or superior to a rating of 4 or any of the other ratings. Ordinal scaling is stronger than nominal scaling, but it is still weak. We cannot meaningfully talk about differences of ratings in an ordinal scale because there are many ways to choose such numbers and differences depend on which numbers are chosen. In question 10 the ratings could just as well have been 20, 10, 5, 2, -7 or A, B, C, D, F. In the social and behavioral sciences such factors as power, prestige, and emotional stability might be measured on an ordinal scale.

> 2. The **ordinal** measurement level classifies data into non-overlapping exhaustive categories, as does the nominal scale, but categories that can be ranked or ordered. The ordinal rankings may or may not involve the use of numbers.

Suppose we obtain the information from five students who had Professor X for statistics shown in Table 11.2.

Table 11.2

Rating	No. of Responses
20	0
10	2
5	1
2	1
-7	1

20 corresponds to Outstanding, 10 to Good, 5 to Satisfactory, 2 to Poor, and -7 to Terrible. If we average the results cited in Table 2, we obtain 4, which does not correspond to any of the ratings available. How is 4 to be interpreted? We know that a class consensus value of 4, as defined by averaging the ratings, puts Professor X between Satisfactory and Poor, which is all that can be said. Because differences between ratings have no meaning, we cannot meaningfully talk about how much below satisfactory and above Poor the consensus value 4 places Professor X.

> **Ordinal Scale:** Values are assigned, at our discretion, to establish a ranking of categories. Differences between rank values have no meaning and because of this an arithmetic average of rank values which differs from the rank values has no clear interpretive meaning.

The data arising from questions 5 and 9 are said to be measured on an **interval scale**. Interval data are more than rankings; they are numerical values for which differences are meaningful. There is a meaningful difference of one degree between each unit. On the Fahrenheit temperature scale the difference between the freezing point (32°F) and boiling point (212°F) of water is divided into 180 intervals representing equal amounts of heat, so that it takes the same amount of fuel to raise the temperature of a fixed quantity of water 1F degree on the scale. Differences are meaningful because of the consistency in the behavior of nature; it's not a matter open to our discretion.

One property lacking with an interval scale is that there is no natural, meaningful zero in the situation. The 0°F mark, for example, is an artificially chosen value. It does not signify no heat, heat being a property of molecular motion. 0°F does not signify absence of molecular motion.

As a result, ratios have no meaning in an interval scale. It is meaningless to say, for example, that an object with a temperature of 60°F is twice as hot as one with a temperature of 30°F.

3.	The **interval** measurement level employs numbers for which differences are meaningful. One can talk about how much more or less of the characteristic the quantity has. There is no meaningful zero in the situation, however.

The data arising from questions 3 and 8 are said to be measured on a **ratio scale**. Ratio scale data have the properties of interval scale data and in addition there is a natural, zero point in the situation so that it is meaningful to consider ratios of measurements. A distance of 4 miles, for example, is twice the distance of 2 miles; an age of 30 years is three times the age of 10 years.

4.	The **ratio** measurement level employs numbers, has the properties of the interval measurement level, and there is a meaningful zero in the situation so that ratios can be formed.

In order of strength, we have the nominal, ordinal, interval, and ratio scales.

11.3 FOOD FOR THOUGHT

1. **A Rating Situation**

Nine students in an introductory psychology course were asked to express their satisfaction with their instructor's teaching quality in terms of the ratings 20 (Outstanding), 10 (Good), 5 (Satisfactory), 2 (Poor), −7 (Terrible). The results obtained are shown in Table 11.3.

Table 11.3

Bob S.	2	Ken M.	2	Nancy R.	5
June K.	20	Amy N.	10	Irene S.	2
Adam G.	5	Bill B.	−7	Vera N.	10

Determine the instructor's mean rating and interpret the result.

2. **The Afore Rating Situation**

Determine the instructor's median rating, interpret the result, and compare these with the mean rating and its interpretation stated in the afore.

3. **I've Twice the Aptitude for College as You?**

   Bob received 1200 on an SAT exam, while Joe scored 600. Bob then turned to his friend and proclaimed: "I've twice the aptitude for college as you." Is this a meaningful statement?

4. **I've Twice as Much Money as You?**

   Joe found that he had $20 while Bob had only $10. He then proclaimed: "I've twice as much money as you." Is this a meaningful statement?

5. **Is He 2.17 Times as Evil as She?**

   In a *New York Post* online poll on who was the most evil person of the millennium, Adolf Hitler came in first with 8.67% of the vote and Hillary Clinton came in sixth with 3.99% of the vote. A. Soltis, "Post Readers: Hitler Was Most Evil," *The New York Post*, Nov. 17, 1999; 2.

   Does this mean, according to the poll, that Adolf Hitler is 2.17 times as evil as Hillary Clinton?

6. **Winner of the Slippery Statistics Achievement Award?**

A panel of five members of the Slippery Statistics Society is to rank the achievements of nominees recommended for the Slippery Statistics Achievement Award. The ranking values available are 100 (Outstanding), 80 (Superb), 70 (Very Slippery), 50 (Slippery). Determination of the Award's recipient has come down to two candidates, Joseph S. (who received rankings 50, 70, 70, 80, 100) and Vera S. (who received rankings 50, 70, 80, 80, 80).

   (a) Should Joseph S. receive the SSA Award because the mean of his rankings, 74, exceeds that of Vera S., 72?

   (b) If your answer to (a) is no, then on what basis would you determine the recipient of the SSA Award?

11.4 Answers/Discussion of Food for Thought

1. **A Rating Situation**

   Determine the instructor's mean rating and interpret the result.

Let us first observe that the ratings are ordinal data and that the mean is not meaningful for ordinal data; not meaningful in what sense? one might ask.

There is no law which prohibits us from determining the mean of numbers on hand; the question is, what sense can we make of the value? Adding up the ratings and dividing by 9 yields 5.4, which does not correspond to any of the ratings. Can 5.4 be interpreted as a consensus value of the class, as defined by the mean, which places their instructor above Satisfactory but below Good? So far, okay; it's when we attempt to go further and draw a conclusion about how much better than Satisfactory and how short of Good the class consensus rates their instructor that we enter the realm of the meaningless. This is because differences are not defined when ordinal data are employed.

2. **The Afore Rating Situation**

   Determine the instructor's median rating, interpret the result, and compare these with the mean rating and its interpretation stated in the afore.

Arranging the 9 ratings in increasing order yields:

$$-7, 2, 2, 2, 5, 5, 10, 10, 20$$

The median is the fifth value, 5; half of the students judge the instructor's teaching as satisfactory or below satisfactory; half judge his teaching as satisfactory or better than satisfactory.

There is no question about the interpretation of the median of ordinal data being meaningful. This cannot be said about the mean of ordinal data.

> It is worthy of note that while the median is independent of the rating system used, this is not the case with the mean.

3. **I've Twice the Aptitude for College as You?**

    Is this a meaningful statement?

No; ratios are not meaningful for interval data.

4. **I've Twice as Much Money as You?**

    Is this a meaningful statement.

Yes; the data are ratio data.

5. **Is He 2.17 Times as Evil as She?**

    Is Adolf Hitler 2.17 times as evil as Hillary Clinton, according to the Poll?

Not meaningful for ordinal data.

6. **Winner of the Slippery Statistics Achievement Award?**

    (a) Joseph S. since the mean of his rankings, 74, exceeds that of Vera S., 72?

    (b) If no to (a), then on what basis should the recipient be determined?

(a) No; the rankings are ordinal data. There is no law which prevents us from computing a mean value of the ranking values given, but the meaning of such a value is open to question since differences of rankings are not defined. Although 74 > 72, there is no basis for concluding that 74 is superior to 72.

(b) Medians are meaningful for ordinal data. Vera S. has a median ranking of 80, which means that two panel members ranked her achievements as Superb or better and two ranked her achievements as Superb or less than Superb. Joseph S. has a median ranking of 70, which means that two panel members ranked his achievements as Very Slippery or better and two ranked his achievements as Very Slippery or less. Since Superb is a higher rank than Very Slippery, the SSA Award should go to Vera S. on the basis of her higher ranking median value.

12

A Role for Mathematical Modeling in Teaching Basic Statistics ? Yes, Most Important.

12.1 PREFACE

Hypothesis testing provides us with a suitable setting for discussion of mathematical modeling, a setting requiring that close attention be paid to the realism of the underlying assumptions/postulates as well as the strings attached.

Two models, 1 and 2, are developed. Model 2 takes into account confounding factors. In teaching basic statistics time pressure rarely allows for developing Model 2, but I introduce confounding factors and point out that there is more to the story than Model 1.

Attention is given to the Consumer Price Index (CPI), the CPI model that underlies it, and the controversy that surrounds the Model's assumptions/postulates.

12.2 MODEL 1: TESTING FOR THE DIFFERENCE BETWEEN MEANS, SMALL SAMPLES, INDEPENDENTLY CHOSEN, APPROACH

> Tests of differences between means where small samples n_1 and n_2 ($n_1 < 30$, $n_2 < 30$) are drawn from populations P_1 and P_2 come with strong statistical strings attached.
>
> 1. It is required that the populations P_1 and P_2 be normal/robust.
> 2. The populations P_1 and P_2 must have the same standard deviations; $\sigma_1 = \sigma_2$.
> 3. The samples must be randomly and independently drawn from P_1 and P_2. That is, the drawing of either sample must not affect the composition of the other.

The test statistic that emerges in this case is the t-statistic

$$t = \frac{\overline{x}_1 - \overline{x}_2}{\sqrt{\frac{(n_1-1)s_1^2 + (n_2-1)s_2^2}{n_1 + n_2 - 2} \cdot \left(\frac{1}{n_1} + \frac{1}{n_2}\right)}}$$

or equivalently,

$$t = \frac{\overline{x}_1 - \overline{x}_2}{\sqrt{\frac{\sum(x_1 - \overline{x}_1)^2 + \sum(x_2 - \overline{x}_2)^2}{n_1 + n_2 - 2} \cdot \left(\frac{1}{n_1} + \frac{1}{n_2}\right)}}$$

with $n_1 + n_2 - 2$ degrees of freedom.

Here $\Sigma(x_1 - \overline{x}_1)^2$ is the sum of the squared deviations from the mean $\overline{x}_1$ of the sample values chosen from population P_1 and $\Sigma(x_2 - \overline{x}_2)^2$ is the sum of the squared deviations from the mean $\overline{x}_2$ of the sample values chosen from population P_2.

12.3 FOOD FOR THOUGHT

1. **Lee Tires (Model 1)**

   The quality control manager of Lee Tires, Janet Lee, claims there is no difference in the average life of the tires made at its Wells and Jason City plants. Two populations emerge, the population of tire lifetimes of the tires made at the Wells plant and the population of the tire lifetimes of the tires made at the Jason City plant, with means μ_1 and μ_2. Janet believes that $\mu_1 - \mu_2 = 0$, or equivalently, $\mu_1 = \mu_2$. To test this hypothesis samples would have to be drawn at random from the two plants and their means $\bar{x}_1$ and $\bar{x}_2$ calculated. If their difference were significantly small, then the null hypothesis of no difference in the population means would be accepted or judgment reserved; if their difference were significantly large, then the null hypothesis of no difference would be rejected and a suitable alternative hypothesis accepted.

   Janet would like to test the null hypothesis $H_o: \mu_1 - \mu_2 = 0$ against the alternative hypothesis $H_a: \mu_1 - \mu_2 \neq 0$ at the 0.05 level on a regular basis. Testing on a regular basis by use of large samples becomes expensive and its cost could quickly exceed the amount budgeted for this purpose. To help keep costs within budget Janet proposes to use small samples of size 11 chosen randomly and independently from the Wells and Jason City plants. The questions that arise are:

   (a) What concerns must be addressed?

   (b) How should we proceed?

   (c) What is the result?

   (d) What is the significance of the result for Lee Tires?

2. **Carl Cairns' Fertilizer Problem (Model 1)**

   To determine whether a new fertilizer is more effective than one in current use for growing corn, Carl Cairns selected 5 test plots on his farm at random which he treated with the fertilizer in current use and another 5 plots at random which he treated with the new fertilizer. The 0.05 level was to be used in carrying out the analysis. The following results (in bushels per acre) were obtained.

Current fertilizer: 70.5, 80.2, 90.3, 60.2, 85.4
New Fertilizer: 73.2, 82.1, 93.1, 63.6, 87.1

(a) Formulate null and alternative hypotheses.

(b) What is the basis for your alternative hypothesis?

(c) What accept/reserve judgment and reject regions emerge?

(d) Are there strings attached that must be addressed?

(e) Assuming that the concerns noted in (d) are satisfied, what would you conclude if you were Carl?

3. **Longlife vs. Neverdie**

A fierce competition has developed between the manufacturers of the Longlife and Neverdie batteries. The Xxon Company, which makes Longlife, was planning to run a series of television commercials claiming that the life of Longlife was unsurpassed by that of Neverdie. To be sure that is was on firm ground it planned to carry out a hypothesis test on the difference between the mean lifetimes of Longlife and Neverdie, $\mu_1 - \mu_2$. The test was to be carried out at the 0.01 level and samples of 13 batteries were to be chosen at random from stocks of Longlife and Neverdie batteries and analyzed.

(a) Set up null and alternative hypotheses.

(b) What is the basis for your alternative hypothesis?

(c) What accept/reserve judgment and reject regions emerge?

(d) The samples drawn yielded the following results. Longlife: $\bar{x}_1 = 42$ (hours), $s_1^2 = 3.4$ (hours); Neverdie: $\bar{x}_2 = 44$ (hours), $s_2^2 = 14.2$ (hours). What do you conclude?

(e) What does your finding mean to the Xxon Company?

(f) The Xxon Company went ahead with its advertising campaign and the manufacturer of Neverdie brought false advertising charges against Xxon to the Fairness in Advertising Association (FAA).

A hearing was subsequently scheduled at which the following exchange took place between an FAA official and an Xxon Company representative.

FAA: Since the mean life of Neverdie exceeds the mean life of Longlife in your test, I conclude that your commercial misrepresents the facts.

Xxon: No; this result does not imply that our claim is untrue. At the 0.01 level of significance the difference in sample means is not statistically significant. The significance level 0.01 is so small that the odds are 99 to 1 that our claim is true.

FAA: In carrying out your hypothesis test, isn't it required that the populations you are sampling from be normally/robust distributed?

Xxon: No; it follows from the Central Limit Theorem that the probability behavior of the difference in sample means does not depend on whether the populations being sampled from are normal/robust.

FAA: I have been informed that a hypothesis test established that the standard deviations of the population you sampled from are unequal. What are the implications of this result for your study?

Xxon: It's not relevant to the study and conclusion reached.

Does the FAA official have cause to take issue with any of the Xxon representative's replies? If yes, what should be said?

4. **Neverdie's Counterattack (Model 1)**

True to its name, Neverdie has launched a counterattack with extensive ads claiming that the life of Neverdie is longer than that of Longlife. To test this claim an independent testing agency chose 6 items at random from a list of battery operated items of different models and makes and operated them by using Neverdie batteries. Another 6 randomly chosen items were operated by using Longlife batteries. The following lifetime data (in hours) were obtained:

Neverdie: 53, 48.2, 38.6, 40.1, 54.2, 42.6
Longlife: 52.2, 47.4, 37.8, 39.4, 53.5, 41.8

(a) Formulate null and alternative hypotheses.

(b) Are there strings attached that must be addressed?

(c) Assuming that the strings attached noted in (b) are satisfied, what conclusion do you reach after carrying out the test at the 0.05 level?

(d) What is the significance of your results to Neverdie?

12.4 Answers/Discussion of Food for Thought

1. **Lee Tires (Model 1)**

(a) Are the populations of tire life times of the tires made at the Wells and Jason City plants normal/robust? Are the population standard deviations σ_1 and σ_2 equal? Are the samples chosen from the two plants independently chosen and chosen at random?

Subject to YES on these questions, we may proceed.

(b) We have H_o, H_a, α and accept/reserve judgment and reject bounds shown in Figure 12.1

$$H_o: \mu_1 - \mu_2 = 0$$

$$H_a: \mu_1 - \mu_2 \neq 0$$

$$\alpha = 0.05$$

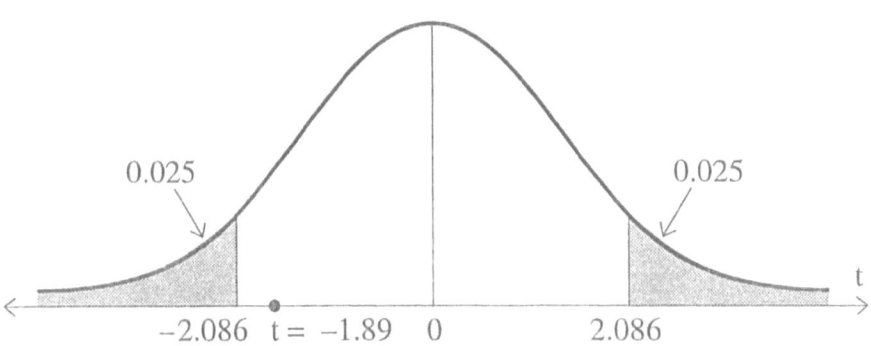

Figure 12.1

Randomly and independently chosen samples of 11 tires from the Wells and Jason City plants yielded the results summarized in Table 12.1.

Table 12.1

	Wells	Jason City
Pop. mean	μ_1	μ_2
Pop. variance	σ_1^2	σ_2^2
Sample size	$n_1 = 11$	$n_2 = 11$
Sample mean	$\bar{x}_1 = 24.6$	$\bar{x}_2 = 26.4$
Sample variance	$s_1^2 = 5.8$	$s_2^2 = 4.2$

(c) The *t*-statistic takes the value

$$t = \frac{24.6 - 26.4}{\sqrt{\frac{10(5.8)+10(4.2)}{20}\left(\frac{2}{11}\right)}} = -1.89,$$

which falls within the accept/reserve judgment region determined by ±2.086 (see Figure 12.1).

(d) The result supports Janet Lee's claim that there is no difference in the average life of the tires made at the Wells and Jason City plants

2. **Carl's Fertilizer Problem (Model 1)**

Let μ_1 and μ_2 denote the average corn yields of plots treated with the fertilizer in current use and the new fertilizer, respectively.

(a) We have:

$H_o: \mu_1 - \mu_2 = 0$

$H_a: \mu_1 - \mu_2 < 0$, or $\mu_1 < \mu_2$

$\alpha = 0.05$

(b) H_a states that use of the new fertilizer results in higher average corn yields (μ_2) than use of the fertilizer currently employed (μ_1).

(c) Since H_a is one of the less than form we need one lower bound to make precise what significantly less than means in a statistical sense. The statistic to be employed is,

$$t = \frac{\overline{X}_1 - \overline{X}_2}{\sqrt{\left(\frac{4s_1^2 + 4s_2^2}{8}\right)\left(\frac{1}{5}+\frac{1}{5}\right)}}$$

The accept/reserve judgment and reject regions that emerge from the lower bound of the t-curve are shown in Figure 12.2.

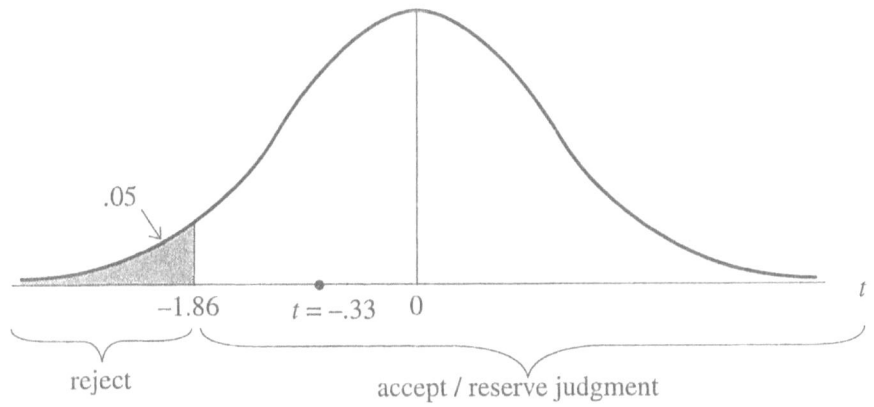

Figure 12.2

(d) Three concerns emerge:

(1) **Normality.** The populations of corn yields obtained through use of the fertilizer currently employed and the new fertilizer must be normal/robust. It is shown that these conditions are satisfied by employing a non-parametric test (see Chapter 2).

(2) **Equality of population standard deviations.** This is shown by the hypothesis test discussed in Sections 1.12 and 1.13. There is a concern about the possibility of committing a Type II error.

(3) **Randomness and Independence of the Samples Chosen.** From the way in which the samples are chosen it is clear that the intent is to satisfy these conditions.

Since the conditions required for carrying out this hypothesis test are satisfied, Carl has a green light to proceed.

(e) For the fertilizer in current use we have:

$$\sum x_1 = 386.3, \quad \sum x_1^2 = 30,474, \quad \bar{x}_1 = 77.3, \quad s_1^2 = 145.4$$

For the new fertilizer we have:

$$\sum x_2 = 399.1, \quad \sum x_2^2 = 32,398, \quad \bar{x}_2 = 79.8, \quad s_2^2 = 135.4$$

This yields:

$$t = \frac{77.3 - 79.8}{\sqrt{\left(\frac{4(145.4) + 4(135.4)}{8}\right)\left(\frac{2}{5}\right)}} = -0.33,$$

which falls in the accept/reserve judgment region (see Figure 12.2).

What this result means to Carl is that there is no advantage to changing fertilizers in terms of increasing average corn yields.

3. **Longlife vs. Neverdie (Model 1)**

(a) $H_o: \mu_1 - \mu_2 = 0$. $H_a: \mu_1 - \mu_2 < 0$.

(b) It contradicts Longlife's claim.

(c) $t = \dfrac{(\bar{x}_1 - \bar{x}_2)}{\sqrt{\left(\dfrac{12s_1^2 + 12s_2^2}{24}\right)\left(\dfrac{1}{13} + \dfrac{1}{13}\right)}}$.

Accept/r. j. on H_o if $t \geq -2.492$. Reject if $t < -2.492$.

(d) $t = -1.72$. Accept/r. j. on H_o.

(e) Longlife's claim is not refuted.

(f) Yes, because of the small sample sizes we must be concerned with the normality/robust of the underlying populations and we must have equality of their variances.

4. **Neverdie Counterattacks (Model 1)**

(a) $H_o: \mu_1 - \mu_2 = 0$. $H_a: \mu_1 - \mu_2 > 0$.

(b) Normality/robust and equality of variances of the underlying populations

(c) $\Sigma x_1 = 276.7$, $\Sigma x^2 = 12{,}982.61$, $\bar{x}_1 = 46.12$, $s_1 = 44.4$.

$\Sigma x_2 = 272.1$, $\Sigma x^2 = 12{,}562.39$, $\bar{x}_2 = 45.45$, $s_2 = 44$.

$t = 0.2 \leq 1.812$. Accept/r. j. on H_o.

(d) Their claim is unsupported by the test.

12.5 Model 2: Test Statistic for The Matched Pairs' Mean Difference Approach

Given the two underlying populations P_1 and P_2 from which the samples are drawn (the populations of corn yields obtained from using the current and new fertilizers, in our case) another approach is to introduce the population of differences P_d obtained by subtracting values y_i of P_2 from the values of x_i of P_1 and examining the behavior of the t-statistic,

$$t = \frac{\bar{x}_d - \mu_d}{s_d / \sqrt{n}}, \quad d.f. = n-1$$

where $\bar{x}_d$, μ_d, s_d and n are the mean of the sample drawn from P_d, mean of P_d, standard deviation of the sample of paired differences and sample size of the sample drawn from P_d, respectively.

$$\bar{x}_d = \frac{\sum(x_i - y_i)}{n}$$

where $x_i - y_i$ is the difference between the paired sample values obtained from P_1 and P_2.

$$\mu_d = \mu_1 - \mu_2,$$

where μ_1 and μ_2 are the means of P_1 and P_2. With respect to the null hypothesis, $\mu_d = 0$:

$$S_d = \frac{\sum(x_i - y_i - \bar{x}_d)^2}{n-1}$$

> The string attached for the application of this result is that the population of differences be normal/robust. If the populations P_1 and P_2 from which P_d is obtained are normal/robust. P_d will be normal/robust.

Model 2: Matched Pairs' Mean Difference for Carl's Fertilizer Problem

> Is Model 1 the best experimental design for extracting information about the effectiveness of the new fertilizer relative to the one in current use? Would it not be better to select 5 plots at random, treat half of each plot with the currently used fertilizer and the other half with the new, and consider the results obtained from the pairings? This approach would more effectively screen out the effects of such extraneous factors as soil conditions, moisture, and weather and focus on the effects of the fertilizer.
>
> The methods of Model 1 are not applicable to paired samples because the string attached that the samples be independently chosen is violated.

Suppose that the resulting yields obtained from the currently used and new fertilizers were obtained from paired plots as summarized in Table 12.2. What conclusion can be drawn at the 0.05 level of significance?

Table 12.2

Pair	Yield x_i (current fertilizer)	Yield y_i (new fertilizer)
1	70.5	73.2
2	80.2	82.1
3	90.3	93.1
4	60.2	63.6
5	85.4	87.1

$$H_o : \mu_d = 0$$

$$H_a : \mu_d < 0$$

$$\alpha = 0.05$$

The accept/r.j. and reject regions for d.f. = 4 are shown in Figure 12.3

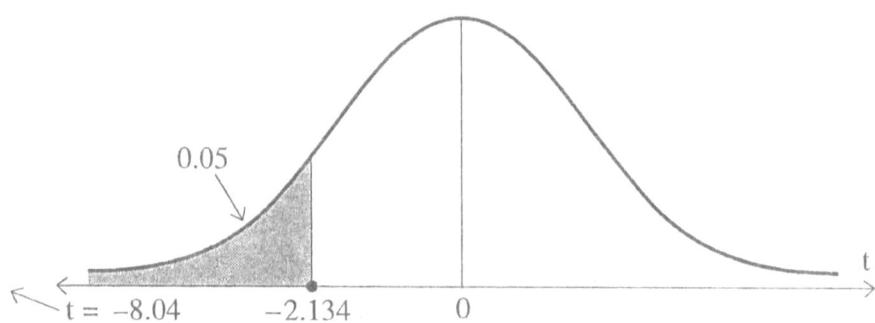

Figure 12.3

The computations for the value of the *t*-statistic come out of Table 12.3

Table 12.3

x_i	y_i	$d_i = x_i - y_i$	$d_i - \bar{x}_d \ (\bar{x}_d = -2.5)$	$(d_i + 2.5)^2$
70.5	73.2	−2.7	−0.2	0.04
80.2	82.1	−1.19	0.6	0.36
90.3	93.1	−2.8	−0.3	0.09
60.2	63.6	−3.4	−0.9	0.81
85.4	87.1	−1.7	0.8	0.64
		−12.5	0	1.94

$$\bar{x}_d = \frac{-12.5}{5} = -2.5, \quad s_d^2 = \frac{1.94}{4} = 0.484, \quad s_d = 0.696$$

$$t = \frac{\bar{x}_d - \mu_d}{s_d / \sqrt{n}} = \frac{-2.5 - 0}{0.696 / \sqrt{5}} = -8.04$$

Since $t = -8.04$ is less than the *t*-bound −2.134, we reject the null hypothesis of no difference and conclude that the new fertilizer is more effective than the one in current use for growing corn.

We should keep in mind that this analysis assumes that the underlying populations of corn yields obtained by use of both fertilizers are normal/robust. This is established in Sec. 2.3 by the Lilliefors test.

It is clear in this situation that by employing matched pair samples we were successful in blocking out extraneous factors and more sharply focusing on the effects of the fertilizers on crop yield.

12.6 Food for Thought

1. **Neverdie's Second Counterattack: Model 2, Matched Pairs' Mean Difference**

   Suppose the Neverdie vs. Longlife test were conducted by choosing 6 items at random from a list of battery operated items and testing both batteries in each of the items. The same data, let us assume, are obtained for the battery lifetimes (hours), but now we have paired values per item as shown in Table 12.4.

   Table 12.4

Item	1	2	3	4	5	6
Neverdie	53	48.2	38.6	40.1	54.2	42.6
Longlife	52.2	47.4	37.8	39.4	53.5	41.8

   (a) Set up null and alternative hypotheses.

   (b) Are there special conditions of concern in carrying out the test?

   (c) Test H_0 at the 0.05 level; what do you conclude?

   (d) What does your finding mean to Neverdie?

   (e) How do you explain the different conclusions obtained here and from Model 1, Sec. 11.3, nu. 4 and 11.4, nu. 4?

   (f) Which approach to the Neverdie versus Longlife test situation is preferable?

2. **Lister Hospital**

   Lister Hospital is considering two vendors, the Veronika and Sabrina Companies, as sources for its medical supplies and drugs.

Ten items were selected from a list of commonly used supplies and medications and their prices (in hundred dollars per unit) for the two vendors noted. The data are summarized in Table 12.5.

Table 12.5

	1	2	3	4	5	6	7	8	9	10
Veronika	2	5.2	10	32	6	11	4	48	15	8
Sabrina	1.5	4.8	11.8	34	5.6	12.7	3.6	50.5	16.7	9

(a) Does the data indicate, at the 0.05 level, that the difference in prices charged by the two vendors is, on average, not the same?

(b) In deciding on which vendor to order supplies from what other factors would you take into consideration?

(c) Another approach to testing for mean price differences is to proceed as follows: Take a random sample of items of interest available from the Veronika Company and determine their mean price $\bar{x}$; independently, take a random sample of items of interest available from the Sabrina Company and determine their mean price $\bar{y}$; by employing the statistic $\bar{x} - \bar{y}$ test the null hypothesis of no difference between the price means of the two vendors against the alternative hypothesis that there is a difference.

Is this approach preferable to the paired-difference approach for obtaining relevant information?

(d) Suppose that the aforedescribed procedure were used and the same data, independently obtained, were determined for the Veronika and Sabrina Companies. Does the data indicate, at the 0.05 level, that there is a difference in the mean prices charged by the two vendors?

(e) Do the conclusions obtained from (a) and (d) agree? If not, which is more revealing?

(f) Bill Adams, a student of statistics, took his statistics teacher's caution, Beware the Assumptions, seriously.

Suppose he found that the population of prices of medical supplies carried by the Veronika company was not normal/robust. What impact, if any, would this have on the analysis given in answer to (a) and (d)?

(g) Suppose Bill found that the population of prices of items carried by the Veronika Company was normal/robust. What impact would this have, if any, on the analysis given in answer to (a) and (d)?

12.7 ANSWERS/DISCUSSION OF FOOD FOR THOUGHT

1. **Neverdie's Second Counterattack: Matched Pairs' Mean Difference (Model 2)**

   (a) $H_o : \mu_d = 0$. $H_a : \mu_d > 0$.

   (b) The normal/robust of the population of differences of lifetimes.

   (c) x_d: 0.8, 0.8, 0.8, 0.7, 0.7, 0.8. $\Sigma x_d = 4.6$, $\Sigma x_d^2 = 3.54$, $\bar{x}_d = 0.767$, $s_d = 0.052$.

   $$t = \frac{0.767 - 0}{0.052/\sqrt{6}} = 36.13 > 2.015. \text{ Reject } H_o.$$

   (d) Their claim is supported by the test.

   (e) The pairing eliminated extraneous factors and focused on the consistent better performance of Neverdie.

   (f) Paired; identified the consistent better performance of Neverdie.

2. **Lister Hospital**

   (a) $H_o : \mu_d = 0$. $H_a : \mu_d > 0$.

   x_d: 0.5, 0.4, −1.8, −2.0, 0.4, −1.7, 0.4, −1.7, 0.4, −2.5, −1.7, −1.0. $\Sigma x_d = -9$,

   $\Sigma x_d^2 = 21$, $\bar{x}_d = -0.9$, $s_d = 1.2$.

   $t = -2.37 < -2.262$. Reject H_o.

We conclude that the differences in prices charged by the vendors are, on average, different.

(b) Reliability, delivery speed, etc.

(c) No. The mean $\bar{x}$ of the items selected from the Veronika Company may be based on less expensive (or more expensive) items that make up the sample (such as relatively inexpensive medications—aspirin and the like) while the mean $\bar{y}$ of the items selected from the Sabrina Company may be based on on more expensive (or less expensive) items that make up the sample—X-ray and high tech equipment).

(d) $H_o : \mu_1 - \mu_2 = 0.$ $H_a : \mu_1 - \mu_2 \neq 0.$

$\Sigma x_1 = 141.2, \Sigma x_1^2 = 3,921.04, \bar{x}_1 = 14.1, s_1 = 214.1.$

$\Sigma x_2 = 150.2, \Sigma x_2^2 = 4,436.28, \bar{x}_2 = 15, s_2 = 242.3.$

$t = -0.13 > -2.101.$ Accept/r.j. H_o.

(e) Disagree; paired differences is better since its focus is on item by item comparison of the two vendor's prices and filters out the afore factor.

(f) It would render the analysis given in answer to (f) and (g) worthless since one string attached for both analyses is that the afore discussed population he normal/robust.

(g) It is a step in the right direction, but two strings attached remain—the normality/robust of the population of prices of items carried by the Sabrina Company for (a) and the afore and equality of population standard deviations for (d).

12.8 THE CONSUMER PRICE INDEX (CPI) MODEL

12.8.1 Preface

In the sample of current basic statistics texts that I examined this topic received shortshrift or no mention. This reflects a serious gap in these texts because this topic affords us an opportunity to discuss a model whose index, published the third week of each month by the Bureau of Labor Statistics (BLS), affects the life of the nation and of every American, directly or indirectly.

In terms of conveying a sense of the importance of statistics to the study of real-world situations, to overlook the CPI Model is tantamount to overlooking the filetmignon before you on your dinner plate

12.8.2 Ripples

The CPI has a wide reach and if it required a theme song my choice would be the one that goes" The knee bone is connected to the thighbone is connected to the hipbone is connected to . . .

Indexed Programs. These are programs for which automatic increases in benefits are triggered by increases in the CPI. Social Security received by nearly 50 million beneficiaries, is the best known of these programs. Others include railroad retirement, with about 800,000 beneficiaries, supplemental Social Security income, with about 6.5 million recipients; veterans' compensation; and Federal military and civilian employee pensions, paid to about 4 million retirees. The official poverty line rises each year in accord with the behavior of the CPI, which affects about 26 million recipients of food stamps, 25 million in subsidized child nutrition programs, and 5 million with federal student grants.

Many workers have union collective bargaining contracts that provide for automatic wage increases when the CPI value rises by a specified amount. The CPI may also be used to adjust such things as alimony and child support payments, attorney's fees, worker compensation payments, apartment, home and office rentals, and welfare payments. In addition to the preceding, movement of the CPI influences Congress on setting the personal tax exemption level and thereby touches most of us in some way.

Taxes. To protect taxpayers from the effects of inflation, taxes are adjusted in a number of ways. This includes tax brackets, which determine tax rates on income; personal exemption and standard deduction levels; earned income tax credit; limit on itemized deductions; and pension contribution limits.

Economic Statistics. Real growth in the Gross Domestic Product (GDP) and productivity (i.e. growth adjusted for inflation) depends on the CPI.

The smaller the increase in the CPI, the smaller will be the additional benefits paid to beneficiaries, the smaller will be the cost to the government, the greater will be the tax obligation of taxpayers and government tax revenue, and the larger will be the GDP and productivity figures.

The Purchasing Power of the Dollar. If comparisons of a dollar's worth over time are to be meaningful, we must have a mechanism for adjusting dollar values to reflect "real buying power" compared to some suitable base used as a point of reference. Price indexes provide such a mechanism.

2.8.3 A Political Dimension

How "solid" is the CPI published by the BLS? As a valid conclusion of its Model, no question about it. As to being a realistic measure of inflation—the subject at the top of our interest list—this received much public attention in the mid - 1990's. Considering the climate for cutting Federal Government spending we can safely bet on its return.

The Essentials of the Story.

In 1995 center stage of government business was taken up by the problem of eliminating the Federal budget deficit. As part of their plan to accomplish this, House Republicans recommended that the annual-cost-of-living adjustment for Social Security and other benefits tied to increases in the Consumer Price Index (CPI) be reduced starting in 1999 [R. Pear; 1]. Washington wisdom circulating at the time held that the CPI overstated inflation by as much as 1.5 percentage points and that a reduction in the CPI's value was not only justified, but defensible.

The problem was to give legitimacy to the Washington wisdom. Since the Bureau of Labor Statistics (BLS) moved cautiously and, from the point of view of the Washington establishment, unreliably in this matter, in June 1995 the Senate Finance Committee appointed a five-member panel of economists, chaired by former President Bush economic advisor Michael Boskin, to study the CPI and make recommendations on revisions. All members of what came to be called the Boskin Commission had respectable credentials and some might be described as eminent, but all had previously given Congressional testimony that the CPI exaggerated inflation. Economists who took a different view, such as former Commissioner of BLS Janet Norwood, were not invited to join the panel.

The Boskin Commission released its report in early December 1996, claiming the consumer inflation was being overstated by the CPI by about 1.1% a year, arguing that the index did not adequately reflect the improving quality of goods, did not take into account new products quickly enough, did not properly reflect consumers' tendency to purchase cheaper alternatives when the price of goods rose, and did not properly take into account consumer shift toward discount stores ([2]. [J. Madrick; 3]).

A number of questions arise: What does it mean to say that the CPI overstates the reality of inflation by about 1.1% a year? Many interpret this to mean that there is an ideal standard for measuring the reality of inflation which is known by the Boskin Commission and that in comparing the BLS's CPI against this ideal standard, the BLS's CPI overstated inflation by about 1.1% a year. This is utter nonsense; there is no ideal standard. The BLS's CPI is a valid conclusion of a math model based on data, accepted procedures, and assumptions, and assumptions made by the agency's economists. The Boskin's Commission's proposed 1.1% per annum adjustment is based on the same data, same procedures, but with somewhat different assumptions. In effect, what the Boskin Commission was saying was that if you employ our assumptions rather than yours, then you have to make a 1.1% per annum downward adjustment in your CPI value. If your CPI increased by 3.3% over 1995, then 2.2% would be a more accurate description of the reality of inflation over that year, based on our assumptions.

Did the Boskin Commission consider the possibility that the BLS's CPI understates inflation? No; for discussion of this situation see [4] to [7].

Is there good reason to prefer the Boskin Commission's assumptions over those made by the BLS? If politics were put aside, it becomes "experts" versus "experts." At a panel session at the annual meeting of the American Economic Association held in New Orleans in January 1997 Boskin and his four commission colleagues presented their views followed by BLS Commissioner Katherine Abraham who stated that 'she agreed with some of the Boskin Commission's recommendations including that the CPI should be as close to a measure of the cost of living as possible.' She added, however, that her agency would not and should not produce a CPI based partly on subjective judgments [J. Berry; 8]. Abraham later further elaborated to the Senate Finance Committee: 'If we get into the business of making judgments about things that are not measurable—guessing, even if it's . . . a best guess—we really, I think, would be undermining the credibility of all of the data we produce. [J. Berry; 8].

The planting of an idea that developed into a drum roll to reduce the CPI was probably done by the Congressional Budget Office when it asserted in late 1994 that the CPI exaggerated inflation by an amount between 0.2 and 0.8 percentage points a year. Federal Reserve Board head Alan Greenspan expressed the view that the CPI exaggerated inflation by an amount between 0.5 and 1.5 percentage points a year at a joint meeting of the House and Senate Budget Committee in January 1995. Greenspan also noted that correcting these estimates could save the government $150 billion over five years and suggested the possibility that Congress pass a law that would lower the CPI by a percentage point or half a percentage point for determining benefits tied to the CPI. It's a short hop from this plateau to the establishment of the Boskin Commission and what became the "official" view that the CPI overstated inflation by about 1.1 percentage points a year.

The Boskin Commission's report set the drum roll into motion in the form of a rash of calls to "fix" the way inflation is measured. Testifying before the Senate Finance Committee on January 30, 1997, Alan Greenspan recommended that an independent commission be established to set cost-of-living adjustments for federal receipts and outlays each year. Economist Martin Feldstein, who had been President Reagan's top economic advisor, suggested that Greenspan's proposed committee should recommend an "appropriate" inflation adjustment factor through informed judgment, apart from any adjustment made to the CPI by the Bureau of Labor Statistics through its normal work. Senators William Roth and Daniel Patrick Moynihan introduced a sense-of-the-Senate resolution that urged an accurate cost-of-living index. With momentum at a peak to push through a CPI fix, it all came apart. President Clinton, faced by strong opposition within his own party and constituencies like Labor and the elderly, decided not to pursue a CPI fix outside of the highly professional, non-political machinery of the Bureau of Labor Statistics. Republican enthusiasm for a "fix" waned with the discovery of a two-year old White House memorandum on how Democrats could use the issue against Republicans.

Within two years talk of engineering a CPI fix to help eliminate the budget deficit had turned to arguments over what to do with the projected budget surplus.

References

1. R. Pear, "G.O.P. Suggests Smaller Benefit Adjustments," *The New York Times,* May 11, 1995.

2. *"Toward a More Accurate Measure of the Cost of Living,* Final Report to the Senate Finance Committee from the Advisory Commission to Study the Consumer Price Index; Dec. 4, 1996.

3. J. Madrick, "The Cost of Living: A New Myth," *The New York Review of Books,* March 6, 1997. Maddrick presents an informative review of and commentary on the Boskin Commission's report and three other books concerning the CPI.

4. D. Baker, "The Inflated Case Against the CPI," *The American Prospect,* Winter 1996.

5. D. Francis, "Fixing the Inflation Index—But is it Really Broken?," *The Christian Science Monitor,* March 6, 1997.

6. D. Francis, "Poking Holes in the C.P.I. Balloon," *The Christian Science Monitor,* March 14, 1997.

7. P. Passell, "Some Experts Say Inflation is Understated," *The New York Times,* Nov. 6, 1997.

8. J. Berry, "A Numbers Game Played for High Stakes," *The Washington Post National Weekly Edition,* March 17, 1997.

12.8.4 Food for Thought

1. When it was argued that the CPI exaggerated inflation by 1.1 percentage points does this mean that there is an "ideal" standard against which the CPI is being compared and found wanting to the extent of 1.1 percentage points?

2. Is it possible that the CPI underestimates inflation?

12.8.5 Answers/Discussion of Food for Thought

1. No. There is no "ideal" standard against which the CPI was measured and found wanting to the degree of 1.1 percentage points. The Boskin Commission's 1.1 percent figure goes back to a disagreement that the Commission had with the assumptions underlying the CPI that had been adopted by the BLS. If you adopt our assumptions rather than yours, the CPI value obtained would be about 1.1 percentage points less, they were in effect arguing.

2. Yes. Reality is actual inflation and the CPI may underestimate reality, which depends on the realism of the assumptions which underlie the CPI model. For discussion see references [4] to [7].

13

KISS (Keep It Simple Stupid), But Keep It Correct.

13.1 Preface

Two issues arose in my examination of a sample of basic statistics texts. The first concerns a formulation of the null hypothesis for a population mean, the second concerns large vs. small sample size for confidence interval determination and hypothesis testing for a population mean.

13.2 "At Least" and "At Most" Claims

Huxley College's SAT Scores

President G. Marx (of Horsefeathers fame) of Huxley College claims that the mean combined SAT scores of Huxley's students is at least 1400. A Higher Education Board (HEB) set up to monitor claims made by colleges and universities to curtail misleading or fraudulent advertising has decided to investigate Huxley's claim.

The first task of HEB's analysts is to set up a null hypothesis. This situation is unlike the others considered in that it is not being claimed that the mean is exactly 1400, but at least 1400, which translates to greater than or equal to 1400. Let us denote the claim by H_c, so that we have:

$$H_c: \mu \geq 1400$$

> We must go from H_c to H_o which must be stated in equality form if we are to emerge with an unequivocal test statistic and probability of a Type I error for a specified level of significance. The underlying theorem is that the mean of the sampling distribution of the mean, $\mu_{\bar{x}}$, equals the mean of the population, μ.

We handle the "at least" condition leading to $H_c: \mu \geq 1400$ in the following way in setting up H_o and H_a.

$$H_o : \mu = 1400$$

$$H_a : \mu < 1400$$

The H_a part is easy; we are led to $\mu < 1400$ as the alternative to the "at least" condition in Huxley's claim that the mean SAT score is at least 1400.

Sample size: $n = 70$.

Level of significance: $\alpha = 0.01$.

Accept/reserve judgment and reject regions: There is one lower bound, $z = -2.33$, to make precise what it means to say that the z-value of $\bar{x}$ is significantly small at the 0.01 level of significance. See Figure 13.1.

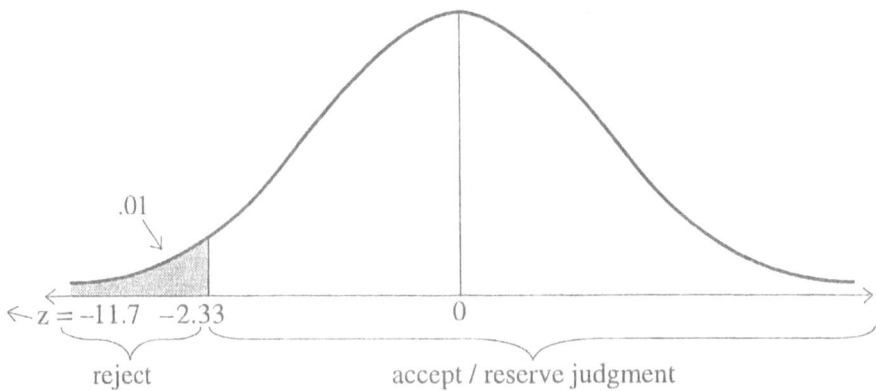

Figure 13.1

The sample of size 70 chosen at random from the population of SAT scores yielded $\bar{x} = 980$ (points) with $s = 300$ (points). The z-value of $\bar{x}$ is

$$z = \frac{980 - 1400}{300/\sqrt{70}} = -11.7,$$

which falls in the reject region.

Pointing to the result obtained, HEB took issue with President Marx's assertion that the mean combined SAT scores of Huxley's students is at least 1400.

An "at least" claim that a population mean μ is at least equal to a value μ_o can be summarized by the condition

$$H_c: \mu \geq \mu_o$$

which in turn may be formulated in terms of the null and alternative hypotheses as follows:

$$H_o: \mu = \mu_o$$

$$H_a: \mu < \mu_o$$

An "at most" claim that a population mean μ is at most equal to a value μ_o can be summarized by the condition

$$H_c: \mu \leq \mu_o$$

which in turn may be formulated in terms of the null and alternative hypotheses in the following way:

$$H_o: \mu = \mu_o$$

$$H_a: \mu > \mu_o$$

13.3 An Incorrect Formulation of the Null Hypothesis for the "At Least" Claim

For Huxley College's SAT Scores the claim $H_c: \mu \geq 1400$ would be taken as the null hypothesis

$$H_o: \mu \geq 1400$$

with

$$H_a: \mu < 1400$$

as the alternative hypothesis.

The analysis would continue as presented in Sec 13.2 with the same conclusion being reached.

The underlying basic theorem

$$\mu_{\bar{x}} = \mu$$

is ignored and an unequivocal test statistic and the probability of a Type I error for the chosen level of significance do not exist.

From an academic who presents such faulty analysis in a published statistics text the consequences are miseducation of students who come in contact with the text.

13.4 Large Sample Size, N=30, Irrespective of the Nature of the Population?

Central Limit Theorem for $\bar{x}$: If a sample of size n is chosen at random from a large population Q (finite or infinite with finite variance), then as n gets larger and larger without bound, the sampling distribution of $\bar{x}$, $p(x) = P(\bar{x} = x)$ gets closer and closer to the normal curve with mean $\mu_{\bar{x}}$ and standard deviation $\sigma_{\bar{x}}$ defined by

$$\mu_{\bar{x}} = \mu, \quad \sigma_{\bar{x}} = \frac{\sigma}{\sqrt{n}},$$

where μ and σ are the mean and standard deviation of population Q.

This raises an important question: If we wish to approximate the sampling distribution of $\bar{x}$ by the normal curve with mean μ and standard deviation $\sigma/\sqrt{n}$, how large must n be for the approximation to be "good enough" for statistical applications?

Theory does not give us an answer to this question.

When I was studying statistics many moons ago the statistics text I used stated that on the basis of statistical practice, n = 30 is taken as a guideline value to define large sample size and that for n ≥ 30 we may use the normal distribution with mean μ and standard deviation $\sigma/\sqrt{n}$ to approximate the sampling distribution of the mean, irrespective of the nature of the underlying population.

The sample of basic statistics books I examined stated a number of points of view which come down to the following:

One text said nothing about how large the sample size must be, but all of the examples presented had n > 30.

A second text said, "as long as n is large (generally n ≥ 30) . . .".

A third text said, "we urge you to be cautious about the sample size. In many practical situations a sample size of n ≥ 30 may be sufficiently large to allow us to use the normal distribution as an approximation for the sampling distribution of the mean. If a population is quite non-normal, the sampling distribution will also be non-normal—even for moderately large values of n."

This point of view has my vote.

14

In Teaching Basic Statistics "Realistic" Data Should Be Used, or Not Necessarily?

14.1 Preface

Not necessarily; in my view realistic situations should be employed when feasible and the data used should be appropriate for the situations.

My primary concern is with focus on basic ideas. I would not use data from real world situations that detract from this primary concern. Moreover, for data that arise from real-world situations, always to be considered is whether the statistics procedure to be employed is suitable for the data.

14.2 Food for Thought

1. Ecap University's School for Statistical Studies

Concerning the situation, names have been changed to protect the guilty.

The Dean of the School for Statistical Studies used the following system to determine merit increments for the school's faculty.

Step 1. Each department chairperson rates each of his/her faculty on Teaching, Scholarship and Research, and Service. The following performance values are to be used:

    5 for Outstanding

    4 or 4.5 for Very Good

    3 or 3.5 for Good

2 or 2.5 for Needs Improvement

1 for Unacceptable

An overall rating is to be obtained by determining the mean of the three ratings.

Step 2. Each chairperson's assessment is reviewed by the Dean, who may change the chairperson's assessment by lowering or raising his/her value assigned.

Step 3. Merit increments (percentage of salary) are determined on the basis of the overall rating as follows:

2.9 or below, 0

3.0 –3.4, 2.8

3.5 –3.9, 3

4.0 –4.4, 3.5

4.5 –4.9, 4

5, 5.5

(a) Is it meaningful to use the mean of the performance ratings?

(b) The Dean lowered the rating assigned by the chairperson of a faculty member for Scholarship and Research from 3 to 2.5. The chairperson informed the faculty member of this result and that his/her overall rating had been lowered to 1. 975.
Is this value meaningful.

14.3 ANSWERS/DISCUSSION OF FOOD FOR THOUGHT

1(a). Is it meaningful to use the mean of the performance ratings?

No. The data are ordinal data and mean values have no meaning for ordinal data. It's the median value which has meaning for ordinal data.

For discussion see Sec. 11.3 and 11.4, Examples 1 and 2.

1(b). Is 1.975 meaningful?

No. Citing 1.975 as the mean value is not meaningful from two points of view. The first is per the discussion of 1(a); the second is, should you accept 1.975 there is the matter of round off to be consistent with the number of decimal places (see Ch. 9).

15

I Believe That Computers and Statistical Packages Should Not Be Used in Teaching Basic Statistics: Heretic, Lunatic, Luddite, or Not?

15.1 PREFACE

At Pace University the Math Department's MAT 117 is a 4-credit basic statistics course required of all BBA students whose major is in the Lubin School of Business. A course in computer programming is not a prerequisite for MAT 117 so that computer background, if needed, must be introduced in the course. Computer usage in MAT 117 is not mandatory, but the pressure was on to make it so.

15.2 W.J. Adams et al; Concerning the Use of Computers and Statistical Packages in a Basic Level Statistics Course

I hope you find the views on the afore useful food-for-thought. Names of colleagues that occur are replaced by W, X, Y, and Z.

October 19, 2009

To: Dr. W., Pace NY Mathematics Department
xc: Pace NY Full-time Mathematics Faculty

From: Prof. W.J. Adams

Subj: **Observations Concerning MAT 117 per Your Memos of Oct. 12 and 13, 2009**

1. The Lubin School has its own priorities concerning the teaching of statistics and from what I see their attitude is that if they were teaching the course, they would do it differently. Not surprising, to say the least.
   While we have been willing to consider their concerns, we have full responsibility for determining priorities. To suggest that the Lubin School is an equal partner in this enterprise, as your delegation of math faculty to "work with Lubin to design a course . . ." certainly suggests, is, in my view, a serious error in judgment.

2. If there is any topic that warrants a department meeting, this one is it. Why the headlong plunge to "start integrating the new syllabus into our classes starting Spring '10" when the "new syllabus" has not been formulated and thoroughly discussed as to its being (1) desirable and (2) feasible? This, in my view, is blantly irresponsible.

3. Imposing the use of statistical software in a basic level course such as MAT 117 is counterproductive to focusing on basic concepts, which should be our first priority. It supports an attitude that we

want to combat—throw your data into the computer and let it do its thing; don't concern yourself with the underlying mathematical hypothesis and the issue of whether the data are reliable and relevant to the enterprise under study.

Furthermore, I believe it's healthy for students to get their hands dirty with computation work. In basic statistics I want my students to compute standard deviation, to take an example, by carrying the computation steps through; it gives them a better sense of the definition of standard deviation than what they would obtain from a statistical package.

I doubt that the Lubin School folks are sensitive to such matters.

Oct. 26, 2009

To: Pace N.Y. Full-time Mathematics Colleagues (except Adams)

From: Z

Dear Colleagues,

Independent of what direction and outcome of the movement of introducing statistical software into MAT 117, I urge you to carefully read and think about the comments that Bill Adams sent to us in his October 19 memo on grading and related manners. There is much wisdom in his words.

Best wishes,
Z

Oct. 27, 2009

To: Z
xc: Full-time Mathematics Colleagues (except Adams)

From: X

If we don't require the use of statistical software, we will not be teaching a statistics course that will be taken by Lubin students. No one—including the Lubin school—is saying we should eliminate all hand computations. The argument about the use of statistical software in statistics courses is silly. At one time people argued over the use of calculators. Computers are tools. We should use them where it is helpful to do so. Students need to be told of their strengths and weaknesses. We can teach a better statistics course with a computer, than without. What we need to do is learn how to incorporate the technology to make a better course. We won't learn how to do it overnight. It will push some of us, perhaps all of us, past our comfort zone. If we do not learn and progress we will become irrelevant. Either we teach statistics with computers or we will not have statistics courses to teach. We ask our students to be life-long learners. It is time we all practice what we preach.

Dec. 7, 2009

To: Colleagues Who Have an Interest in Teaching Elementary Statistics (MAT 117, for example)

From: Prof. William J. Adams

Subj: **Statistical Packages for Teaching Elementary Statistics? NO.**

Prof. W circulated an e-mail memo that encourages the use of a statistical package in MAT 117 beginning with the Spring 2010 semester. I disagree with my learned colleague, which prompts this rejoinder.

Commandment 7 of Adams-Kabus-Preiss, *Statistics*: *Basic Principles and Applications*. "The computation dimension should not be given equal billing with statistical principles and ideas." "Statistics is the master and, important as it is, the computation tool is the servant."

How so?

1. **Counter-productive**

    The use of statistical packages in a basic level statistics course is counter-productive to what should be the course's first priority—placing the focus on basic concepts, theorems, and conditions under which they are applicable. Not enough attention is now being given to conditions under which theorems are applicable and the use of statistical packages further obscures what should be our principal focus.

2. **Go with the Flow?**

    The direction of the flow these days is to throw your data into the computer and let it do its thing.

    Corollary 1: Mathematical Hypotheses

    Don't concern yourself with underlying mathematical hypotheses. (Indeed, what do you mean by such?)
    It is my view that this flow must be fought in order for sound statistics education to prevail. The use of statistical packages in a basic level statistics course reinforces the strength of this flow.

    Corollary 2: Reliability and Relevance of Data

    With numbers, data, and statistics being hurled at us on an ongoing basis this dimension is more essential to sound statistics education than ever before. We should enhance our attention to this dimension in MAT 117, not detract from it.

    Corollary 3: Miraculous Conversion

    Many of our students leave our MAT 117 course with the view that the exercise of statistical techniques convert data into indisputable wisdom. Go with the Flow nurtures this misunderstanding.

3. **Dirty Hands Favor Healthy Understanding**

I believe that, within reason, it's healthy for students to get their hands dirty with computation work. To take an example, in basic statistics I want my students to compute standard deviation by carrying through the computation steps. It gives them a better sense of what standard deviation is about than they would obtain from a statistical package or pressing a button on a programmable calculator.

The Proper Place for Statistical Packages

I favor using statistical packages in statistics courses beyond the basic level.

Dec. 10, 2009

To: Full-time Pace N.Y. Math Colleagues
From: Z

As before I urge you to carefully read Bill Adams' views on Computer Software in our classes. His latest memo is dated December 7.

Do not lose sight of his wisdom on elementary classes.

His point concerning lower level courses and required software should not be ignored.

Having said this, the reality is we have no choice. Simply put, no software means no classes for us. The department has such a weakened position at Pace that we could not get any support if we tried to continue as we have been doing. My own experience has been that the students for the most part absorb very little understanding of the concepts. Unfortunately, this will be even more so as we lose more class time devoted to discussing software. Nevertheless, we most do the best we can with our move to required software.

Dec. 14, 2009

To: Dr. Z, Full-time Math Colleagues

From: Prof. William J. Adams

Subj: Dr. Z's Dec. 10th e-mail, aka I Come to Praise Ceasar AND to Bury him

Faculty Council Curriculum Committee Meeting, Nov. 23, 2009

I attended this meeting chaired by Dr. Y. After we had been serenaded by calls for business applications in our statistics course, I talked about MAT 117, pointing out that it focused on basic concepts, conditions under which theorems and concepts are applicable, and the issue of the reliability and relevance of data that are the basic raw materials. For the sake of honesty I added the qualifier that I probably gave more attention to these concerns than some of my colleagues. I further commented that at this basic level there are two "serious" applications to business and economics—index numbers and time series—and that apart from introductory discussion of these ideas they are best handled by colleagues in the disciplines in which they occur. (Of course we give illustrations with a business flavor of concepts developed in MAT 117, but I distinguish those from applications integral to other disciplines—what I termed "serious" applications.)

I believe that this commentary turned the thrust of the meeting around 180 degrees. Dr. Y, who teaches statistics with a focus on applications to psychology, commented that perhaps they should require MAT 117 as a prerequisite for their statistics course. He reflected what I perceive as a consensus that the math department do the fundamentals (as I described them) in MAT 117 and that they handle "serious" applications from their disciplines. This part of the meeting ended on this note.

Statistics Basics and Us

My sense of the outcome of the meeting is that teaching statistics fundamentals is the math department's responsibility, at least for the time being. The string attached is that we take serious steps to implement on a general scale the focus I described.

The computation *über alles* direction enhanced by statistical packages (our equivalent of the French Maginot Line) makes us copycats of the business statistics mentality. If we reduce our thinking to this level, the basic statistics course could just as well be given by the business school. The scene then reduces to only a math turf vs. business turf issue.

15.3 EXCEL FANS: CAUTION

"The statistical features within Excel were not originally part of the software, . . . recently statisticians have discovered certain issues with some of the statistical procedures, some of these issues are minor aggravations . . . Others are considered more serious, under certain circumstances the output results are incorrect and do not match what you would get from true statistical software packages such as Minitab.[1] *Issues with Excel* are discussed on pp. 31, 48, 91, 93, 354.

Issues with Excel are not taken up in the sixth edition of [1] and I asked a colleague (June 2011) who is knowledgeable about this issue whether this means that the *Issues with Excel* have been corrected. Not to his knowledge, he told me, and he follows such matters closely.

---

[1] D. Groebuer, P. Shannon, P. Fry, K. Smith; *Basic Statistics: A Decision Making Approach*, 5th ed. (Prentice Hall, 2001), p. 31.

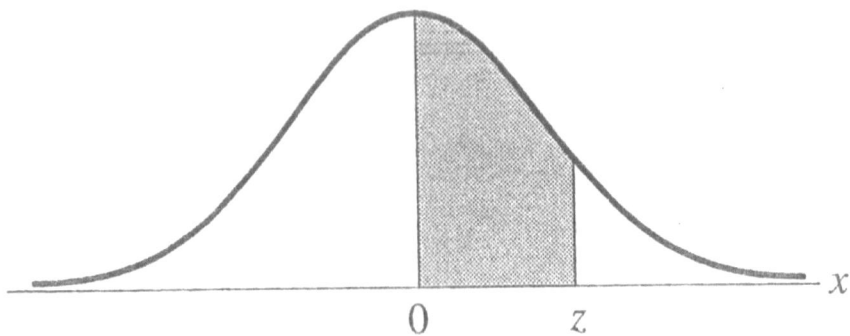

Table A
Standard Normal Curve

z	.00	.01	.02	.03	.04	.05	.06	.07	.08	.09
.0	.0000	.0040	.0080	.0120	.0160	.0199	.0239	.0279	.0319	.0359
.1	.0398	.0438	.0478	.0517	.0557	.0596	.0636	.0675	.0714	.0753
.2	.0793	.0832	.0871	.0910	.0948	.0987	.1026	.1064	.1103	.1141
.3	.1179	.1217	.1255	.1293	.1331	.1368	.1406	.1443	.1480	.1517
.4	.1554	.1591	.1628	.1664	.1700	.1736	.1772	.1808	.1844	.1879
.5	.1915	.1950	.1985	.2019	.2054	.2088	.2123	.2157	.2190	.2224
.6	.2257	.2291	.2324	.2357	.2389	.2422	.2454	.2486	.2518	.2549
.7	.2580	.2612	.2642	.2673	.2704	.2734	.2764	.2794	.2823	.2852
.8	.2881	.2910	.2939	.2967	.2995	.3023	.3051	.3078	.3106	.3133
.9	.3159	.3186	.3212	.3238	.3264	.3289	.3315	.3340	.3365	.3389
1.0	.3413	.3438	.3461	.3485	.3508	.3531	.3554	.3577	.3599	.3621
1.1	.3643	.3665	.3686	.3708	.3729	.3749	.3770	.3790	.3810	.3830
1.2	.3849	.3869	.3888	.3907	.3925	.3944	.3962	.3980	.3997	.4015
1.3	.4032	.4049	.4066	.4082	.4099	.4115	.4131	.4147	.4162	.4177
1.4	.4192	.4207	.4222	.4236	.4251	.4265	.4279	.4292	.4306	.4319
1.5	.4332	.4345	.4357	.4370	.4382	.4394	.4406	.4418	.4429	.4441
1.6	.4452	.4463	.4474	.4484	.4495	.4505	.4515	.4525	.4535	.4545
1.7	.4554	.4564	.4573	.4582	.4591	.4599	.4608	.4616	.4625	.4633

1.8	.4641	.4649	.4656	.4664	.4671	.4678	.4686	.4693	.4699	.4706
1.9	.4713	.4719	.4726	.4732	.4738	.4744	.4750	.4756	.4761	.4767
2.0	.4772	.4778	.4783	.4788	.4793	.4798	.4803	.4808	.4812	.4817
2.1	.4821	.4826	.4830	.4834	.4838	.4842	.4846	.4850	.4854	.4857
2.2	.4861	.4864	.4868	.4871	.4875	.4878	.4881	.4884	.4887	.4890
2.3	.4893	.4896	.4898	.4901	.4904	.4906	.4909	.4911	.4913	.4916
2.4	.4918	.4920	.4922	.4925	.4927	.4929	.4931	.4932	.4934	.4936
2.5	.4938	.4940	.4941	.4943	.4945	.4946	.4948	.4949	.4951	.4952
2.6	.4953	.4955	.4956	.4957	.4959	.4960	.4961	.4962	.4963	.4964
2.7	.4965	.4966	.4967	.4968	.4969	.4970	.4971	.4972	.4973	.4974
2.8	.4974	.4975	.4976	.4977	.4977	.4978	.4979	.4979	.4980	.4981
2.9	.4981	.4982	.4982	.4983	.4984	.4984	.4985	.4985	.4986	.4986
3.0	.4987	.4987	.4987	.4988	.4988	.4989	.4989	.4989	.4990	.4990

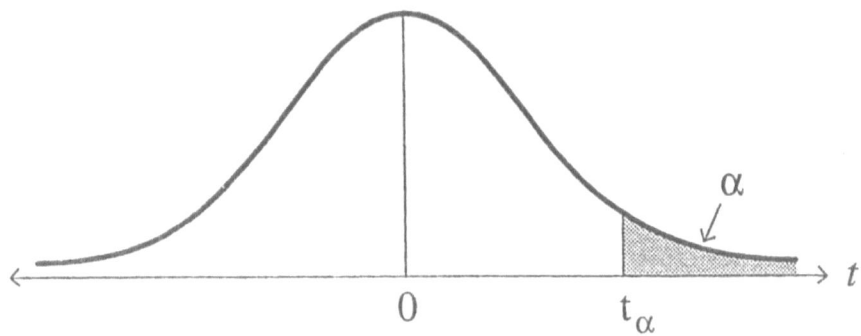

Table B
t Distribution

Degrees of Freedom	$t_{.10}$	$t_{.05}$	$t_{.025}$	$t_{.01}$	$t_{.005}$
1	3.078	6.314	12.706	31.821	63.657
2	1.886	2.920	4.303	6.965	9.925
3	1.638	2.353	3.182	4.541	5.841
4	1.533	2.132	2.776	3.747	4.604
5	1.476	2.015	2.571	3.365	4.032
6	1.440	1.943	2.447	3.143	3.707
7	1.415	1.895	2.365	2.998	3.499
8	1.397	1.860	2.306	2.896	3.355
9	1.383	1.833	2.262	2.821	3.250
10	1.372	1.812	2.228	2.764	3.169
11	1.363	1.796	2.201	2.718	3.106
12	1.356	1.782	2.179	2.681	3.055
13	1.350	1.771	2.160	2.650	3.012
14	1.345	1.761	2.145	2.624	2.977
15	1.341	1.753	2.131	2.602	2.947
16	1.337	1.746	2.120	2.583	2.921

17	1.333	1.740	2.110	2.567	2.898
18	1.330	1.734	2.101	2.552	2.878
19	1.328	1.729	2.093	2.539	2.861
20	1.325	1.725	2.086	2.528	2.845
21	1.323	1.721	2.080	2.518	2.831
22	1.321	1.717	2.074	2.508	2.819
23	1.319	1.714	2.069	2.500	2.807
24	1.318	1.711	2.064	2.492	2.797
25	1.316	1.708	2.060	2.485	2.787
26	1.315	1.706	2.056	2.479	2.779
27	1.314	1.703	2.052	2.473	2.771
28	1.313	1.701	2.048	2.467	2.763
29	1.311	1.699	2.045	2.462	2.756
30	1.310	1.697	2.042	2.457	2.750
40	1.303	1.684	2.021	2.423	2.704
60	1.296	1.671	2.000	2.390	2.660
120	1.289	1.658	1.980	2.358	2.617

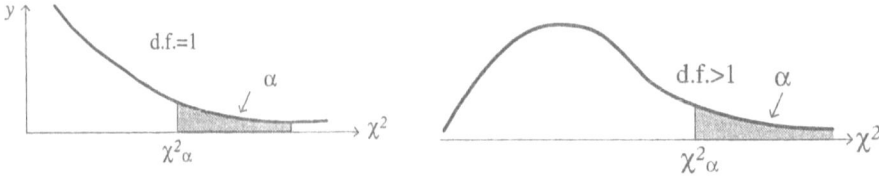

Table C
Chi-Square Curve

d.f.	$\chi^2_{.995}$	$\chi^2_{.99}$	$\chi^2_{.975}$	$\chi^2_{.95}$	$\chi^2_{.90}$	$\chi^2_{.10}$	$\chi^2_{.05}$	$\chi^2_{.025}$	$\chi^2_{.01}$	$\chi^2_{.005}$
1	.0000393	.000157	.000982	.00393	.01579	2.706	3.841	5.024	6.635	7.879
2	.0100251	.0201007	.0506356	.102587	.210720	4.60517	5.99147	7.37776	9.21034	10.5966
3	.0717212	.114832	.215795	.351846	.584375	6.25139	7.81473	9.34840	11.3449	12.8381
4	.206990	.297110	.484419	.710721	1.063623	7.77944	9.48773	11.1433	13.2767	14.8602
5	.411740	.554300	.831211	1.145476	1.61031	9.23635	11.0705	12.8325	15.0863	16.7496
6	.675727	.872085	1.237347	1.63539	2.20413	10.6446	12.5916	14.4494	16.8119	18.5476
7	.989265	1.239043	1.68987	2.16735	2.83311	12.0170	14.0671	16.0128	18.4753	20.2777
8	1.344419	1.646482	2.17973	2.73264	3.48954	13.3616	15.5073	17.5346	20.0902	21.9550
9	1.734926	2.087912	2.70039	3.32511	4.16816	14.6837	16.9190	19.0228	21.6660	23.5893
10	2.15585	2.55821	3.24697	3.94030	4.86518	15.9871	18.3070	20.4831	23.2093	25.1882
11	2.60321	3.05347	3.81575	4.57481	5.57779	17.2750	19.6751	21.9200	24.7250	26.7569
12	3.07382	3.57056	4.40379	5.22603	6.30380	18.5494	21.0261	23.3367	26.2170	28.2995
13	3.56503	4.10691	5.00874	5.89186	7.04150	19.8119	22.3621	24.7356	27.6883	29.8194
14	4.07468	4.66043	5.62872	6.57063	7.78953	21.0642	23.6848	26.1190	29.1413	31.3193
15	4.60094	5.22935	6.26214	7.26094	8.54675	22.3072	24.9958	27.4884	30.5779	32.8013
16	5.14224	5.81221	6.90766	7.96164	9.31223	23.5418	26.2962	28.8454	31.9999	34.2672
17	5.69724	6.40776	7.56418	8.67176	10.0852	24.7690	27.5871	30.1910	33.4087	35.7185
18	6.26481	7.01491	8.23075	9.39046	10.8649	25.9894	28.8693	31.5264	34.8053	37.1564
19	6.84398	7.63273	8.90655	10.1170	11.6509	27.2036	30.1435	32.8523	36.1908	38.5822
20	7.43386	8.26040	9.59083	10.8508	12.4426	28.4120	31.4104	34.1696	37.5662	39.9968
21	8.03366	8.89720	10.28293	11.5913	13.2396	29.6151	32.6705	35.4789	38.9321	41.4010
22	8.64272	9.54249	10.9823	12.3380	14.0415	30.8133	33.9244	36.7807	40.2894	42.7958
23	9.26042	10.19567	11.6885	13.0905	14.8479	32.0069	35.1725	38.0757	41.6384	44.1813
24	9.88623	10.8564	12.4011	13.8484	15.6587	33.1963	36.4151	39.3641	42.9798	45.5585
25	10.5197	11.5240	13.1197	14.6114	16.4734	34.3816	37.6525	40.6465	44.3141	46.9278

26	11.1603	12.1981	13.8439	15.3791	17.2919	35.5631	38.8852	41.9232	45.6417	48.2899
27	11.8076	12.8786	14.5733	16.1513	18.1138	36.7412	40.1133	43.1944	46.9630	49.6449
28	12.4613	13.5648	15.3079	16.9279	18.9392	37.9159	41.3372	44.4607	48.2782	50.9933
29	13.1211	14.2565	16.0471	17.7083	19.7677	39.0875	42.5569	45.7222	49.5879	52.3356
30	13.7867	14.9535	16.7908	18.4926	20.5992	40.2560	43.7729	46.9792	50.8922	53.6720
40	20.7065	22.1643	24.4331	26.5093	29.0505	51.8050	55.7585	59.3417	63.6907	66.7659
50	27.9907	29.7067	32.3574	34.7642	37.6886	63.1671	67.5048	71.4202	76.1539	79.4900
60	35.5346	37.4848	40.4817	43.1879	46.4589	74.3970	79.0819	83.2976	88.3794	91.9517
70	43.2752	45.4418	48.7576	51.7393	55.3290	85.5271	90.5312	95.0231	100.425	104.215
80	51.1720	53.5400	57.1532	60.3915	64.2778	96.5782	101.879	106.629	112.329	116.321
90	59.1963	61.7541	65.6466	69.1260	73.2912	107.565	113.145	118.136	124.116	128.299
100	67.3276	70.0648	74.2219	77.9295	82.3581	118.498	124.342	129.561	135.807	140.169

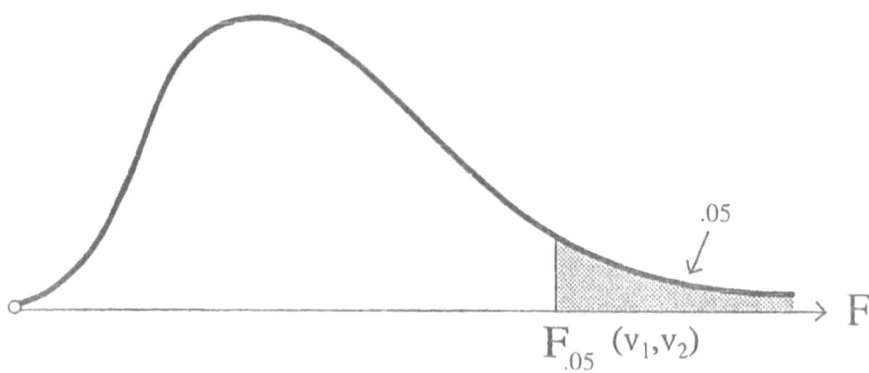

Table D
$F_{.05}(v_1, v_2)$

v_2 \ v_1	1	2	3	4	5	6	7	8	9	10	12	15	20	24	30	40	60
1	161	200	216	225	230	234	237	239	241	242	244	246	248	249	250	251	252
2	18.5	19.0	19.2	19.2	19.3	19.3	19.4	19.4	19.4	19.4	19.4	19.4	19.4	19.5	19.5	19.5	19.5
3	10.1	9.55	9.28	9.12	9.01	8.94	8.89	8.85	8.81	8.79	8.74	8.70	8.66	8.64	8.62	8.59	8.57
4	7.71	6.94	6.59	6.39	6.26	6.16	6.09	6.04	6.00	5.96	5.91	5.86	5.80	5.77	5.75	5.72	5.69
5	6.61	5.79	5.41	5.19	5.05	4.95	4.88	4.82	4.77	4.74	4.68	4.62	4.56	4.53	4.50	4.46	4.43
6	5.99	5.14	4.76	4.53	4.39	4.28	4.21	4.15	4.10	4.06	4.00	3.94	3.87	3.84	3.81	3.77	3.74
7	5.59	4.74	4.35	4.12	3.97	3.87	3.79	3.73	3.68	3.64	3.57	3.51	3.44	3.41	3.38	3.34	3.30
8	5.32	4.46	4.07	3.84	3.69	3.58	3.50	3.44	3.39	3.35	3.28	3.22	3.15	3.12	3.08	3.04	3.01
9	5.12	4.26	3.86	3.63	3.48	3.37	3.29	3.23	3.18	3.14	3.07	3.01	2.94	2.90	2.86	2.83	2.79
10	4.96	4.10	3.71	3.48	3.33	3.22	3.14	3.07	3.02	2.98	2.91	2.85	2.77	2.74	2.70	2.66	2.62
11	4.84	3.98	3.59	3.36	3.20	3.09	3.01	2.95	2.90	2.85	2.79	2.72	2.65	2.61	2.57	2.53	2.49
12	4.75	3.89	3.49	3.26	3.11	3.00	2.91	2.85	2.80	2.75	2.69	2.62	2.54	2.51	2.47	2.43	2.38
13	4.67	3.81	3.41	3.18	3.03	2.92	2.83	2.77	2.71	2.67	2.60	2.53	2.46	2.42	2.38	2.34	2.30
14	4.60	3.74	3.34	3.11	2.96	2.85	2.76	2.70	2.65	2.60	2.53	2.46	2.39	2.35	2.31	2.27	2.22
15	4.54	3.68	3.29	3.06	2.90	2.79	2.71	2.64	2.59	2.54	2.48	2.40	2.33	2.29	2.25	2.20	2.16
16	4.49	3.63	3.24	3.01	2.85	2.74	2.66	2.59	2.54	2.49	2.42	2.35	.228	2.24	2.19	2.15	2.11
17	4.45	3.59	3.20	2.96	2.81	2.70	2.61	2.55	2.49	2.45	2.38	2.31	2.23	2.19	2.15	2.10	2.06
18	4.41	3.55	3.16	2.93	2.77	2.66	2.58	2.51	2.46	2.41	2.34	2.27	2.19	2.15	2.11	2.06	2.02
19	4.38	3.52	3.13	2.90	2.74	2.63	2.54	2.48	2.42	2.38	2.31	2.23	2.16	2.11	2.07	2.03	1.98
20	4.35	3.49	3.10	2.87	2.71	2.60	2.51	2.45	2.39	2.35	2.28	2.20	2.12	2.08	2.04	1.99	1.95
21	4.32	3.47	3.07	2.84	2.68	2.57	2.49	2.42	2.37	2.32	2.25	2.18	2.10	2.05	2.01	1.96	1.92
22	4.30	3.44	3.05	2.82	2.66	2.55	2.46	2.40	2.34	2.30	2.23	2.15	2.07	2.03	1.98	1.94	1.89
23	4.28	3.42	3.03	2.80	2.64	2.53	2.44	2.37	2.32	2.27	2.20	2.13	2.05	2.01	1.96	1.91	1.86
24	4.26	3.40	3.01	2.78	2.62	2.51	2.42	2.36	2.30	2.25	2.18	2.11	2.03	1.98	1.94	1.89	1.84
25	4.24	3.39	2.99	2.76	2.60	2.49	2.40	2.34	2.28	2.24	2.16	2.09	2.01	1.96	1.92	1.87	1.82

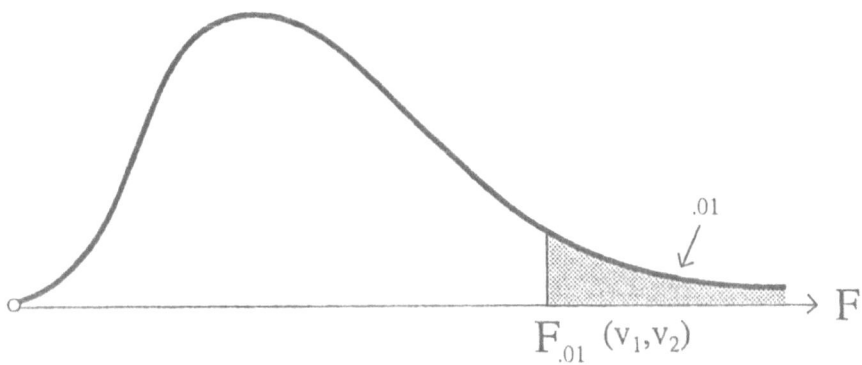

Table E
$F_{.01}(v_1, v_2)$

$v_2 \backslash v_1$	1	2	3	4	5	6	7	8	9	10	12	15	20	24	30	40	60
1	4,052	5,000	5,403	5,625	5,764	5,859	5,928	5,982	6,023	6,056	6,106	6,157	6,209	6,235	6,261	6,287	6,313
2	98.5	99.0	99.2	99.2	99.3	99.3	99.4	99.4	99.4	99.4	99.4	99.4	99.4	99.5	99.5	99.5	99.5
3	34.1	30.8	29.5	28.7	28.2	27.9	27.7	27.5	27.3	27.2	27.1	26.9	26.7	26.6	26.5	26.4	26.3
4	21.2	18.0	16.7	16.0	15.5	15.2	15.0	14.8	14.7	14.5	14.4	14.2	14.0	13.9	13.8	13.7	13.7
5	16.3	13.3	12.1	11.4	11.0	10.7	10.5	10.3	10.2	10.1	9.89	9.72	9.55	9.47	9.38	9.29	9.20
6	13.7	10.9	9.78	9.15	8.75	8.47	8.26	8.10	7.98	7.87	7.72	7.56	7.40	7.31	7.23	7.14	7.06
7	12.2	9.55	8.45	7.85	7.46	7.19	6.99	6.84	6.72	6.62	6.47	6.31	6.16	6.07	5.99	5.91	5.82
8	11.3	8.65	7.59	7.01	6.63	6.37	6.18	6.03	5.91	5.81	5.67	5.52	5.36	5.28	5.20	5.12	5.03
9	10.6	8.02	6.99	6.42	6.06	5.80	5.61	5.47	5.35	5.26	5.11	4.96	4.81	4.73	4.65	4.57	4.48
10	10.0	7.56	6.55	5.99	5.64	5.39	5.20	5.06	4.94	4.85	4.71	4.56	4.41	4.33	4.25	4.17	4.08
11	9.65	7.21	6.22	5.67	5.32	5.07	4.89	4.74	4.63	4.54	4.40	4.25	4.10	4.02	3.94	3.86	3.78
12	9.33	6.93	5.95	5.41	5.06	4.82	4.64	4.50	4.39	4.30	4.16	4.01	3.86	3.78	3.70	3.62	3.54
13	9.07	6.70	5.74	5.21	4.86	4.62	4.44	4.30	4.19	4.10	3.96	3.82	3.66	3.59	3.51	3.43	3.34
14	8.86	6.51	5.56	5.04	4.70	4.46	4.28	4.14	4.03	3.94	3.80	3.66	3.51	3.43	3.35	3.27	3.18
15	8.68	6.36	5.42	4.89	4.56	4.32	4.14	4.00	3.89	3.80	3.67	3.52	3.37	3.29	3.21	3.13	3.05
16	8.53	6.23	5.29	4.77	4.44	4.20	4.03	3.89	3.78	3.69	3.55	3.41	3.26	3.18	3.10	3.02	2.93
17	8.40	6.11	5.19	4.67	4.34	4.10	3.93	3.79	3.68	3.59	3.46	3.31	3.16	3.08	3.00	2.92	2.83
18	8.29	6.01	5.09	4.58	4.25	4.01	3.84	3.71	3.60	3.51	3.37	3.23	3.08	3.00	2.92	2.84	2.75
19	8.19	5.93	5.01	4.50	4.17	3.94	3.77	3.63	3.52	3.43	3.30	3.15	3.00	2.92	2.84	2.76	2.67
20	8.10	5.85	4.94	4.43	4.10	3.87	3.70	3.56	3.46	3.37	3.23	3.09	2.94	2.86	2.78	2.69	2.61
21	8.02	5.78	4.87	4.37	4.04	3.81	3.64	3.51	3.40	3.31	3.17	3.03	2.88	2.80	2.72	2.64	2.55
22	7.95	5.72	4.82	4.31	3.99	3.76	3.59	3.45	3.35	3.26	3.12	2.98	2.83	2.75	2.67	2.58	2.50
23	7.88	5.66	4.76	4.26	3.94	3.71	3.54	3.41	3.30	3.21	3.07	2.93	2.78	2.70	2.62	2.54	2.45
24	7.82	5.61	4.72	4.22	3.90	3.67	3.50	3.36	3.26	3.17	3.03	2.89	2.74	2.66	2.58	2.49	2.40
25	7.77	5.57	4.68	4.18	3.86	3.63	3.46	3.32	3.22	3.13	2.99	2.85	2.70	2.62	2.53	2.45	2.36

Table F
Lilliefors Test Bounds

Sample Size n	Significance Level α				
	.20	.15	.10	.05	.01
4	.300	.319	.352	.381	.417
5	.285	.299	.315	.337	.405
6	.265	.277	.294	.319	.364
7	.247	.258	.276	.300	.348
8	.233	.244	.261	.285	.331
9	.223	.233	.249	.271	.311
10	.215	.224	.239	.258	.294
11	.206	.217	.230	.249	.284
12	.199	.212	.223	.242	.275
13	.190	.202	.214	.234	.268
14	.183	.194	.207	.227	.261
15	.177	.187	.201	.220	.257
16	.173	.182	.195	.213	.250
17	.169	.177	.189	.206	.245
18	.166	.173	.184	.200	.239
19	.163	.169	.179	.195	.235
20	.160	.166	.174	.190	.231
25	.142	.147	.158	.173	.200
30	.131	.136	.144	.161	.187
Over 30	$\dfrac{.736}{\sqrt{n}}$	$\dfrac{.768}{\sqrt{n}}$	$\dfrac{.805}{\sqrt{n}}$	$\dfrac{.886}{\sqrt{n}}$	$\dfrac{.1031}{\sqrt{n}}$

Index

A

Abraham, Katherine, 162
Anderson, Martin, 121

B

Barrett, Laurence, 63
BLS. *See* Bureau of Labor Statistics (BLS)
Boskin, Michael, 161-62
Boskin Commission, 161-64
Bureau of Labor Statistics (BLS), 99, 158-59, 161-64

C

Carter, Jimmy Carter, 63
China, Cultural Revolution of, 66, 111
CIA procedure, major fallacy in, 107
Clinton, Bill, 100, 163
consumer price index (CPI), 98-100, 139-40, 158-64
CONTAM (Committee on Nationwide Television Audience Measurement), 52
Cortines, Ramon, 58
Costikyan, Edward, 59
CPI. *See* consumer price index (CPI)
Cultural Revolution of China. *See* China, Cultural Revolution of

D

D'Amato, Alfonse, 62
Data Scales, 12, 129
Divorce Revolution, The (Weitzman), 102
dollar, purchasing power of, 160
Drolet, Robert, 120
Dukes, Graham, 60

E

economic statistics, 100, 160
Excel, issues with, 181

F

Falkner, David, 119
Last Yankee, The, 119, 124
Feldstein, Martin, 163
Fiddes, Robert, 101
fully indexed programs, 99

G

Gates, Robert, 108
GDP. *See* gross domestic product (GDP)
GIGO (garbage in, garbage out), 58
Goodman, Marvin Goodman, 108
Great Depression, 101
Greenspan, alan, 162-63

gross domestic product (GDP), 83, 100, 160
Grundberg, Ilo, 59

H

Halcion, 59-60
Handbook 8, *61*
Hayes, Rutherford, 104
Higher education Board (HeB), 166-67
Hite, Shere, 111-12
  Hite Report, The, 111-12
  Hite Report on Male Sexuality, The, 111-12
  Women and Love, 111
Hite Report, The (Hite), 111
Hite Report on Male Sexuality, The (Hite), 111
Holzman, Franklyn D., 105, 107-9
Huxley College, 84, 125, 166, 168-69
Hypothesis Testing, 139

I

interval measurement level, 133
interval scale, 132-33

K

Kirichenko, V. N., 68

L

Last Yankee, The (Falkner), 119, 124
Leontief, Wassily, 122, 124
Lilliefors test, 38-39, 190

M

manufacturing output, 81
McMillion, Charles, 80-81, 85
Midwest Economic Development Association (MEDA), 29-30
Moynihan, Daniel Patrick, 163

N

national debt of United States. *See* United States, national debt of
Norwood, Janet, 161
null hypothesis, 43, 165

O

ordinal scale, 131-32
Oswald, Ian, 60

P

paired-sample sign test, 43-44
people-meter data gathering system, 52
poll results, trustworthiness of, 90

R

random sampling, 50, 52
ratio measurement level, 133
ratio scale, 133
RC-1, 85-88
RC-2, 86-88
Reagan, Ronald, 65, 69, 71, 103, 121-24, 163
Reagan Boom, 124
Ricupero, Rubens, 66

robustness, 35
Rome Laboratory, 61
Roth, William, 163
Russell, Michael, 85-86

S

Schulman, Mark A., 97
Slippery Statistics Society (SSS), 66, 135
Stockman, David, 63-65, 71
Strings Attached, 17, 19

T

taxes, 99, 160
Thayer, Frederick C., 100
Thompson, William C., Jr., 59
Tilden, Samuel, 104
TV ratings, 51-52

U

unemployment, 80-81, 121, 123
United States, national debt of, 82
U.S. Competitiveness, 81

V

von Neumann, John, 55

W

Weidenbaum, Murray, 63-64, 103
Weitzman, Lenore, 102
  Divorce Revolution, The, 102
Will, George F., 119
Wisdom, Ivor M., 84
Women and Love (Hite), 111

Z

Zhang Sai, 66

www.ingramcontent.com/pod-product-compliance
Lightning Source LLC
Chambersburg PA
CBHW030939180526
45163CB00002B/638